U0300598

恭祝万达集团
成立 25 周年

2009
万达商业规划
WANDA COMMERCIAL PLANNING

中国建筑工业出版社

北京铂尔曼大饭店

pullman
HOTELS
man
RESORTS
BEIJING WEST WANDA
北京万达铂尔曼大饭

Etam
LOREAL

西安民乐园万达广场中庭

《万达商业规划2009》编委会
THE EDITORIAL BOARD OF WANDA COMMERCIAL PLANNING 2009

WANDA 2009

卓越的规划设计是商业地产的核心武器
EXCELLENT PLANNING AND DESIGN ARE THE KEY WEAPONS OF THE COMMERCIAL REAL ESTATE

万达之所以能进入发展的快车道！核心因素是万达商业模式的成熟……成熟有四方面的标志：

Why Wanda Group can enter the fast lane of development? The core factor is the mature business model··· There are four characteristics of that mature model.

一是形成完整产业链。万达有中国唯一的商业规划院！有专门的建设团队!有中国最大的商业管理团队！能完全独立设计、开发建设、运营管理商业地产。

The first point is to form a complete industrial chain. Wanda Group has the one and only technical management and research institution in China. The Institute has specialized construction team and China's largest commercial management team, which design, develop, operate, and manage commercial real estate independently.

二是商业资源丰富。现在万达已不是招商！而是选商！确保订单地产模式的顺利实施！万达项目在开工时招商就基本完成！开业时保证满场开业……

The second point is to have rich business resources. Now Wanda Group doesn't need to invite merchant, but select merchant! This will ensure the smooth implementation of the order model of the estate. The work of inviting merchant has been done at the beginning of the project construction to ensure the full opening of the retail shops at the Wanda Plaza's Grand-open day.

三是营运能力很强。保证万达广场开一个火一个。万达现在不仅能做到招满商，而且能旺场，这在全国都是奇迹……

The third point is to have a strong operation ability. Now Wanda is able to not only recruit merchants to 100%, but also ensure every Wanda Plaza has a exuberant start. This is a miracle in China.

四是内控制度完善！万达风险控制模式能确保万达拿项目时不犯颠覆性错误，保证后期运作顺利，使项目开发的各项指标全部能控制在目标范围之内。

The fourth point is to have a perfect internal control system! Wanda's risk control mode ensure Wanda does not make disruptive mistakes when taking the project. ensure the smooth operation of the subsequent operations, and ensure all the Project development indicators can be controlled in the target range.

这些都标志着万达商业地产模式的基本成熟，这是十多年不断摸索、学习、总结的结果。

All of these marks Wanda commercial real estate model has been basically mature. This is the result of ten years continuous exploration, learning and summary.

——王健林董事长， 万达集团2009年半年会

Wang Jianlin

Chairman of Board and President of Wanda Group

at 2009 Semiannual Meeting

万达集团董事长

王健林

Wang Jianlin

Chairman of Board and President of Wanda Group

目录
CONTENTS

万达商业规划与万达商业规划研究院

WANDA COMMERCIAL PLANNING AND WANDA COMMERCIAL PLANNING & RESEARCH INSTITUTE

万达集团董事长王健林指导规划院工作
Wang Jianlin guiding Wanda Commercial Planning & Research Institute

■ 万达商业规划是万达商业地产成功模式的重要组成部分。
Wanda commercial planning is an important success element of its commercial real estate business.

万达商业规划的核心规划原则以及业务模式，是万达集团董事长王健林先生亲自指导和创立的。万达商业规划院经过多年的发展，形成的以设计管理为基础、以品质提升为目标，坚持创新发展、绿色节能，全面配合商业地产开发运营的全球独创的业务模式，多次被王健林董事长称为“商业地产创新模式的龙头及重要核心”。万达商业规划在万达商业地产中的作用不仅是规划设计类业务，更是商业地产整体产业链上实现各方业务整合串联的中枢环节。万达商业规划的作用主要表现在以下四个方面。

The core planning principle and business model of Wanda commercial planning is personally directed and created by Chairman Wang Jianlin of Wanda Group. After years of development, Wanda Commercial Planning & Research Institute has formed its business model that is uniquely created in the world, featuring holistic support of the development and operation of commercial real estate, on the basis of designing management and with the goal of enhancing quality, at the same time of adhering to innovative development and green energy saving. It has been termed numerous times by Chairman Wang Jianlin as the "engine and important core of the innovation model of commercial real estate". The role of Wanda commercial planning in Wanda commercial real estate is not only in terms of planning and designing, but also as the central hub for realizing the integration and connection of all different businesses along the entire industry chain of commercial real estate. The role of Wanda commercial planning is mainly manifested in the following four aspects.

1. **关联业务最广**。万达规划类业务横向几乎是关联到了产业链上的所有业务部门，包括发展、设计、工程、成本、招商、计划、安监、质监、营销、财务、法务、企划、人力、运营等，规划设计类业务仅占有限的比例，大部分为规划管理类业务。万达规划管理类业务包括：万达商业地产产品的指标管控、品质监造、成本协调、商户对接、营销配合、标准制定、绿建节能、专利申请、复盘创新、运营跟踪等方面。因此，万达商业规划研究院是万达成熟的“项目计划模块”操作体系中，关联度最广、管控节点占比最高的总部管理部门。

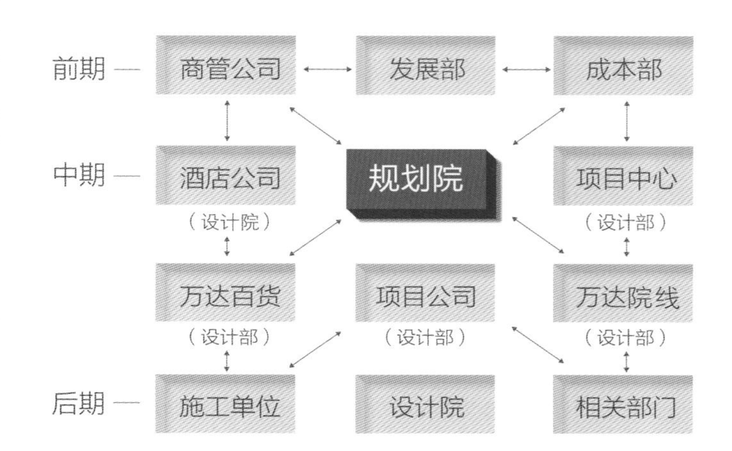

1. It involves the broadest scope of businesses. Horizontally, Wanda planning business almost involves all business sectors along the industry chain, including departments responsible for development, designing, engineering, cost, investment promotion, scheduling, safety inspection, quality inspection, marketing, financial affairs, legal affairs, corporate planning, human resources, operation, etc.. The business core of Wanda Commercial Planning & Research Institute is planning and designing, but it only accounts for a limited proportion. Most of its business is in the category of planning management. Wanda's planning management business includes: the index control, quality supervision, cost coordination, commercial tenants interfacing, marketing support, standards formulation, green building & energy saving, patent application, readjustment innovation, operation tracking and other aspects of the products of Wanda commercial real estate. Therefore, among all management departments at the HQ of Wanda, Wanda Commercial Planning & Research Institute involves the broadest aspects and has the highest proportion of control milestones in the mature "Project Scheduling Module" operating system of Wanda.

2. 管控纵深最长。万达商业规划研究院纵向贯穿项目开发建设及运营全过程：在项目摘牌前阶段，参与选址、规划及地质等风险评判，以及与测算互动的总图及指标调整。在前期设计阶段，完成总图规划的业态布局、指标分配、建造标准的确定；完成方案设计的业态确认、功能分区、动线设计；完成初步设计的整体设计、商户落位，夯实指标及带单成本；完成施工图设计及材料封样把控，消防人防批复。在中期实施阶段，进行招商对接配合、设计调整，进行现场施工封样确认及品质指导管理，进行项目中期检查、安全检查、绿色建筑节能检查和竣工验收。在后期运营阶段，完成项目档案、参与复盘，对产品进行回访调研，确定创新及整改方向。

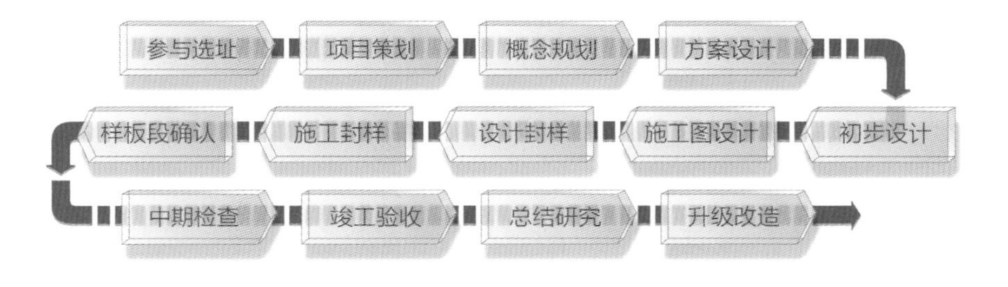

2. It has the longest controlling depth. Vertically, Wanda Commercial Planning & Research Institute penetrates the whole course of project development, construction and operation. In the phase before bid-winning of the project, it participates in the site selection, planning as the risk evaluation of geological factors, etc.. Also it will adjust the master plan and the master plan index interacting with estimation. In the phase of designing, it will decide the layout of various formats, allocation of index and the standards of construction of master-planning; it will also decide the types of businesses, functions zoning and kinetonema designing in the schematic design; it will also finish the overall designing, complete the layout map of commercial tenants and make sure of index and conditional budget; it will also complete the designing of construction drawing, controlling of materials sealing and the acquisition of approvals from fire safety and civil air defense authorities. In the mid-term phase of implementation, it will support the interfacing with commercial tenants, adjust the designing, confirm the sealing at construction site, provide quality instruction and management, conduct mid-term inspection, safety inspection, green building and energy saving inspection and completion acceptance. In the phase of operation, it will finish archives-related work, participate in the readjustment, conduct surveys on the projects and determine the directions for innovation and rectification.

3. 影响成本最大。商业规划设计对商业地产测算及成本的影响大到数以“亿”元计：前期选址及业态功能规划不合理对地产现金流的影响，商业动线设计不合理对后期运营的影响，规划条件如地质、人防、市政、航空限高以及微波通道等未协调落实对后期建造的影响，均是大到数以“亿”元计。在设计阶段，设计对建造工程的影响，对后期租赁平效的长期影响，可以说是以“千万”元计的。在建造阶段，对规划设计的调整如果不加控制，会带来数以“百万”元计的建造成本的变化。因此，“规划创造价值，设计产生效益”是万达商业规划的价值所在。

3. It impacts the largest proportion of cost. The impact of commercial planning and designing on commercial properties estimation and cost is in "hundred millions" yuan, including the impact of the unreasonable preliminary site selection and planning of format functions on the cash flow, the impact of unreasonable designing of commercial kinetonema on the phase of operation and the impact of failure to fulfill geological, civil air defense, municipal, air traffic height limitation and microwave channel requirements, all of which are in "hundred millions" yuan. In the phase of designing, the impact of designing on the construction and the floor effects of leasing in the operation phase is in "ten millions" yuan. In the phase of construction, if the adjustment of planning and designing is not well controlled, changes in construction cost will be in "millions" yuan. Therefore, the value of Wanda commercial planning lies in the following slogan: "Planning creates value, while designing creates profits".

4. 带动创新最多。万达商业规划与万达商业管理的密切对接及复盘，体现了万达商业规划的创新来源于实践，服务于运营“有用”原则。万达商业规划研究院组织对已建成的万达广场和酒店，进行工程建设及管理运营的规划设计复盘，从实践中不断总结经验，并将复盘成果持续应用在新设计的产品中。万达商业规划是万达产品不断创新与发展的动力。

4. It drives the largest proportion of innovation. The close interfacing between Wanda commercial planning and Wanda commercial management as well as readjustment draws the innovation of Wanda commercial planning from practice and makes it serve the operation. Wanda Commercial Planning & Research Institute organizes planning and designing readjustment on Wanda Plazas and hotels that have been built up, in terms of construction, management and operation. Thus the experience can be summarized through practice and the results of readjustment can be applied to the newly designed products continually. Wanda commercial planning is the engine for the continual innovation and development of Wanda's products.

■ 万达商业规划设计的主体是万达广场即万达商业综合体，万达广场的商业规划经历了第一代单店、第二代组合店，形成了第三代万达商业综合体。

The carrier of Wanda commercial planning and designing is Wanda Plaza, namely, Wanda Commercial Complex. The commercial planning of Wanda Plaza has experienced the following periods: the 1st generation featuring single unit, the 2nd generation featuring multiple units and the 3rd generation, Wanda commercial complex.

万达商业综合体是以万达购物中心、万达高星级酒店为核心，以室外商业步行街或商业街为连接，按照业态功能进行分区布局，结合商务办公、居住、教育等多种复合功能的高端成熟的城市综合体。关于万达广场的发展历程、万达广场的总图设计、万达广场购物中心的单体设计、万达酒店的设计、万达商业规划的管控管理，我们将在本次出版的系列年册中进行梗概介绍。

The core of Wanda commercial complex is Wanda shopping center and Wanda high-star hotels, with outdoor commercial pedestrian walks or commercial streets as links. Its zoning layout is based on the functions of its different formats, integrating multiple functions including commercial business, residence and education. Wanda commercial complex is a high-end mature urban complex. We will provide a rough introduction of the development history and general plot plan of Wanda Plaza, the building designing of Wanda Plaza shopping center, the designing of Wanda hotels and the control and management of Wanda commercial planning in this series of yearbook.

哈尔滨中央大街万达广场
Harbin Central Street Wanda Plaza

哈尔滨中央大街万达广场总平面图
The master plan of Harbin Central Street Wanda Plaza

南京新街口万达广场
Nanjing Xinjiekou Wanda Plaza

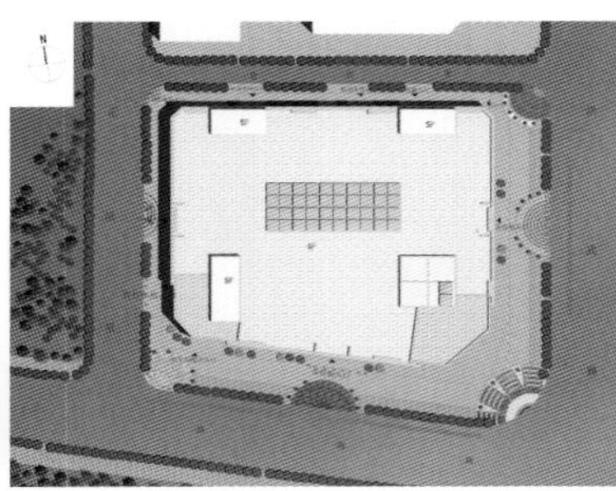
南京新街口万达广场总平面图
The master plan of Nanjing Xinjiekou Wanda Plaza

福州仓山万达广场
Fuzhou Cangshang Wanda Plaza

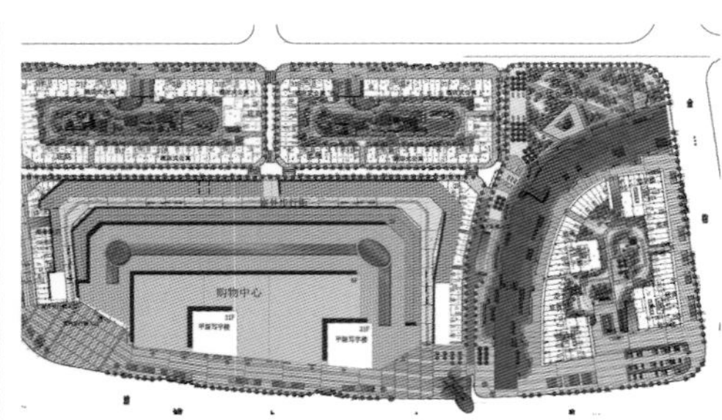

福州仓山万达广场总平面图
The master plan of Fuzhou Cangshang Wanda Plaza

■ 万达商业规划的管理体系，从属并服务于万达商业地产项目开发整体管理体系，实现了“规划设计全过程信息化管控”。“规划设计全过程信息化管控”是在集团“项目信息化管控计划模块”的基础上，建立规划设计图文档系统，与集团信息化办公管理平台联动，形成三大互为关联的信息化操作平台。规划设计管理可以从摘牌开始，按照计划模块展开每个节点，按每个节点完成成果，并对每个节点成果自动归档，直至竣工复盘完成。全程信息化管控体系具有节点提醒、标准查询、成果检索、分类编辑等强大管理功能。万达规划设计的全程信息化管控体系，使万达商业规划管理达到了全球顶级水平。

The management system of Wanda commercial planning belongs to and serves the overall management system of the development of the project of Wanda commercial real estate, materializing the "whole course computerized and informationized control of planning and designing". The "whole course computerized and informationized control of planning and designing" establishes the computer graphics and files system for planning and designing on the basis of the Group's "Project Informationization Control Planning Module". It synergizes with the Group's informationization management platform, forming three interconnected computer informationization operating platforms. The planning and designing management can start from bid-winning and fulfill all tasks in accordance with all milestones in the planning module. The result of each milestone will be automatically archived until the ending of construction and readjustment. The computer control system has strong management functions including milestones reminding, standards inquiry, results retrieval and categorized editing. The whole course computerized and informationized control system of Wanda planning and designing upgrade Wanda commercial planning management to the world's top level.

■ 万达商业规划研究院有限公司（简称万达商业规划研究院）注册成立于2007年。万达商业规划院是万达集团的技术归口管理部门，是万达商业地产模式的核心竞争力之一，也是全球第一家依托房地产公司，兼具设计生产、过程管控、标准研发三大职能，主要从事大型商业建筑、酒店建筑、文化旅游类建筑规划设计及规划设计管控的公司。

Wanda Commercial Planning & Research Institute Co., Ltd. (Wanda Commercial Planning & Research Institute for short) was formally established in 2007. It is Wanda Group's competent department for technical management and is one of the core competitive edges of Wanda's commercial real estate model. It is also the first company in the world dedicated to the planning, designing and controlling of large-scale commercial buildings, hotel buildings and cultural tourism buildings, as one part of a real estate group, with all three major functions of designing & production, process control and standards R&D.

万达商业规划研究院的设计生产职能，主要体现在万达商业地产持有物业的设计上，即负责万达购物中心、高星级酒店、甲级写字楼、商业步行街、剧场等的总图设计、建筑方案设计及初步设计。

Wanda Commercial Planning & Research Institute's designing and production function is mainly reflected in the main designing of the properties owned by Wanda Commercial Properties Co., Ltd. The Institute takes charge of the general layout designing, architectural scheme designing and preliminary designing of Wanda Plazas, luxury hotels, Class A office buildings, commercial pedestrian streets, theatres, and so on.

万达商业规划研究院的设计管控职能，主要体现在负责全部施工图设计的整体协调及管控以及与设计相关的安全、计划、成本、面积指标、产品品质、绿色建筑节能六大方面的管控上。

万达集团执行董事、总裁丁本锡（左二）和大连万达商业地产股份有限公司执行总裁齐界（右一）对规划院工作进行现场指导
The executive director / president of Wanda Group Ding Benxi (L2) and CEO Qi Jie (R1) guiding Wanda Commercial Planning & Research Institute

Wanda Commercial Planning & Research Institute's management and control function is mainly reflected in the coordination and control of the entire construction drawings designing and design-related control in 6 aspects, including safety, schedule, cost, area index, product quality, and green building and energy saving.

万达商业规划研究院的**标准研发职能**，主要体现在代表万达集团参与编制行业规范标准、企业制度标准以及各类专利的申请。万达商业规划研究院成立仅5年，便参与并主要起草了由住房和城乡建设部、商务部、公安部颁布的与商业相关的规范及标准5部，涉及商业建筑、商业管理、消防安全、绿建节能、游乐机械等多个相关领域，申请并获得国际专利20余项。

Wanda Commercial Planning & Research Institute's standard research and development function is mainly reflected in doing work on behalf of the Group, for example: formulation of the industrial norms and standards and corporate systems and standards as well as the application of various kinds of patents. Within 5 years of Wanda Commercial Planning & Research Institute's history, the institute takes part in the drafting of 5 business-related standards for the Ministry of Housing and Urban-Rural Construction, Ministry of Commerce and Ministry of Public Security. The contents involve several related fields in commercial buildings, commercial management, fire safety, green building and energy saving as well as amusement machinery. The institute applied and acquired more than 20 international patents.

■ 万达商业规划研究院的绿建节能研究所是王健林董事长提议设立的，是集团唯一的绿建节能工作整体科研和管理部门。2010年，万达商业规划研究院绿建节能研究所牵头完成了《万达集团“绿色、低碳”战略研究报告》，首次公开明确地提出了万达集团商业建筑“绿色、低碳”战略目标，即：“2011年及以后开业的项目均达到绿色建筑一星设计标准；2011年至2015年间新开业项目逐年降低运行能耗2%～3%；2013年取得5个项目绿色建筑一星运行标识认证；2015年实现运营管理水平全部达到绿色建筑一星运行标准”；截至截稿日，通过住房和城乡建设部绿色建筑设计标识认证的十多个大型商业建筑全是万达广场！在本次出版的项目中，广州白云万达广场被评为绿色建筑二星设计标识，万达学院被评为绿色建筑三星设计标识，均是所在类型建筑目前国内星级最高设计认证标识！2012年，首批共3个万达广场通过了住房和城乡建设部绿色建筑一星运行标识认证，实现了国内大型商业绿色节能运营“零”的突破！

The establishment of Green Building & Energy Conservation Research Department of Wanda Commercial Planning & Research Institute is proposed by Mr. Wang Jianlin, Chairman of Wanda Group. This department is the only holistic research and management department that takes charge of environmental protection and energy saving work within the Group. In 2010, the Department took the lead in finishing the "Green & Low Carbon" Strategic Research Report, which clearly proposed the "Green & Low Carbon" strategic objective of Wanda Group's commercial building publicly for the first time. The strategic objectives are: all projects opened for business in and after 2011 shall reach the designing standard of 1-star green building designing; projects already opened for business between 2011 and 2015 shall reduce their operational energy consumption by 2%~3% each year; 5 projects shall be awarded "1-star green building operation certification label" in 2013; the level of operation and management shall reach "1-star standard for green building operation" in 2015. So far, all of the more than ten large-scale commercial buildings approved by the Ministry of Housing and Urban-Rural Construction are Wanda Plazas. Among these projects, Guangzhou Baiyun Wanda Plaza was given a 2-star Green Building Construction Certification Label and Wanda Institute was given a 3-star Green Building Construction Certification Label. All these are China's highest construction certification label for all kinds of buildings! In 2012, the first batch of three Wanda Plazas received the 1-star Green Building Certification Label issued by the Ministry of Housing and Urban-Rural Construction, which materialized a breakthrough in the field of domestic large-scale commercial green energy saving and operation!

■ 万达商业规划研究院及万达商业规划一直是在王健林董事长的亲自指导和关怀下不断地创新发展；万达商业规划体系的标准化、信息化建设也一直是在丁本锡总裁的直接领导下建立和完善。在此谨对王健林董事长的指导和关怀表示感谢，对丁本锡总裁的领导和重视表示感谢，对万达集团的所有部门和同事的支持表示感谢，并借此机会向支持和帮助过万达商业规划研究院的设计行业的同仁们及所有配合的设计供方表示感谢。

Wanda Commercial Planning & Research Institute and Wanda's commercial planning has always been innovating and developing under the direct instruction and care of Chairman Wang Jianlin. The standardization and informationization of Wanda commercial planning system has been established and refined under the direct leadership of President Ding Benxi all along. We, Wanda Commercial Planning & Research Institute, hereby express our gratitude to Chairman Wang Jianlin's instruction and care, to the leadership and attention of President Ding Benxi and to the support of all other departments and the colleagues of Wanda Group. Also, we would like to take this opportunity to extend our gratitude to those peers in the designing industry and all the other cooperative design companies who have supported and helped us all along.

■ 本次由万达商业规划研究院主编，中国建筑工业出版社出版的《万达商业规划2009》、《万达商业规划2010》及《万达商业规划2011》，是万达商业规划研究院首次向社会出版发行的万达商业地产产品成果介绍专辑。截止到2011年年底，万达集团在全国已开业46个万达广场，24个五星级及五星级以上酒店，自持物业面积达到903万平方米，持有物业规模进入世界前八，是中国民营企业的典范和中国民族企业的骄傲。为使社会更好地了解万达，我们将陆续主编出版万达商业类、文化旅游类规划设计图书，以图册年鉴及专辑这两类主要形式，介绍万达商业及文化旅游项目的规划设计成果并与社会分享。

The books of *Wanda Commercial Planning 2009*, *Wanda Commercial Planning 2010* and *Wanda Commercial Planning 2011* are the first effort of Wanda Commercial Planning & Research Institute to publish the products and achievements of Wanda's commercial real estate to the society. By the end of 2011, Wanda Group has opened 46 Wanda Plazas and 24 hotels of five-star and even above across China. The area of properties owned by Wanda Group reaches 9.03 million square meters and this enables Wanda Group to be among the Top 8 in the world in terms of the scale of self-owned properties. Therefore, Wanda Group can be well termed as an example of China's private companies and the pride of China's national enterprises. In order to make Wanda better known to the society, we will publish books on the introduction of the planning and designing of Wanda's commercial and cultural tourism projects one after another.

《万达商业规划2009》、《万达商业规划2010》及《万达商业规划2011》向大家展示了三年内开业的37个万达广场及23个万达酒店，并在《万达商业规划2009》中对2009年之前开业的10个万达广场及6个酒店做图册性的收录。2013年将出版《万达商业规划2012》，介绍万达2012年开业的18个万达广场、10个五星级及五星级以上酒店，之后每年将陆续出版此类图册年鉴。

Wanda Commercial Planning 2009, *Wanda Commercial Planning 2010* and *Wanda Commercial Planning 2011* display 37 Wanda Plazas and 23 Wanda hotels opened for business within three years. And *Wanda Commercial Planning 2009* collects the photos of the 10 Wanda Plazas and 6 hotels opened for business before 2009. *Wanda Commercial Planning 2012* will be published in 2013, introducing 18 Wanda Plazas and 10 hotels of five-star and even above, all opened for business in 2012. Such pictured yearbooks will be published successively each year in the future.

作为万达商业规划系列年鉴式图册，我们将在每年分册中分别概述万达规划发展不同阶段的内容、标志性事件以及对商业地产发展的作用和贡献，与商业地产同行分享规划设计及其管理相关的发展案例；同时，我们也将在每年分册中陆续发表与商业规划相关的学术及管理型文章，与同行交流，为提高我国商业地产的规划设计及其管理水平尽万达应尽的社会责任；最后，我们也诚挚地希望各界同行及关注万达的朋友们对我们的工作提出宝贵意见！

As a serial yearbook-style atlas for Wanda commercial planning, it will separately describe the contents, landmark events as well as the roles and contributions to the development of commercial real estate, sharing cases of development in planning and designing as well as the management thereof with our peers in the commercial real estate industry. At the same time, we will also publish academic and management articles related to commercial planning in the serial books published each year, to communicate with industry peers. This is also a move to fulfill Wanda's due social responsibilities in enhancing the planning and designing as well as the management of China's commercial real estate. Finally, we also sincerely hope that our peers from all circles and friends paying attention to Wanda can give valuable feedbacks on our work!

大连万达商业地产股份有限公司高级总裁助理
万达商业规划研究院院长
赖建燕
2012.12.30
Lai Jianyan
Senior Assistant of the President of Dalian Wanda Commercial Estate Co., LTD
President of Wanda Commercial Planning & Research Institute
Dec 30, 2012

万达商业规划2009

WANDA COMMERCIAL PLANNING 2009

《万达商业规划2009》中收录了2009年年度开业的8个万达广场、2个五星级酒店，同时也将2007年万达商业规划院成立以来2007年、2008年开业的万达广场及酒店做了年鉴目录性收录。至此，连同以后每年出版的年册，共同组成向社会展示万达商业规划研究院成立以来规划设计的万达商业、酒店类建筑的完整年鉴类档案。

Wanda Commercial Planning 2009 covers 8 Wanda Plazas opened for business in 2009 and 2 five-star hotels. Also, the Wanda Plazas and hotels opened for business in 2007 and 2008, after the establishment of Wanda Commercial Planning & Research Institute, are also recorded as part of the yearbook catalogue. So far, along with yearbooks in future years, we can display to society a complete yearbook-style record of Wanda's commercial buildings and hotels planned and designed since the establishment of Wanda Commercial Planning & Research Institute.

2009年是万达商业规划研究院的发展年。万达自2000年步入商业地产以来到2008年，每年平均的开业数量基本稳定在2～3个万达广场、1～2个五星级酒店。2009年首次快速成倍发展。

2009 is the year of growth for Wanda Commercial Planning & Research Institute. Since Wanda Group entered the field of commercial real estate, the number of projects opened for business each year, from 2002 to 2008, is steady: 2~3 Wanda Plazas and 1~2 five-star hotels. It was in 2009 that the first rapid and exponential growth emerged.

2009年是万达商业地产逆势而上，加大开发力度的第一次提速。2009年当年开业8个万达广场；之后的2010年，开发建设速度再次提升，当年开业万达广场15个，五星级酒店5个，商业及酒店同时增加近一倍；到2011年，开业14个万达广场，12个酒店，酒店开业数量又增加一倍多！万达商业地产经过2009年后连续三年的提速开发建设，到 2012年，当年开业万达广场18个，五星级酒店12个。此后持有物业的发展速度基本稳定在每年开业万达广场17～20个，五星级酒店10～12个。

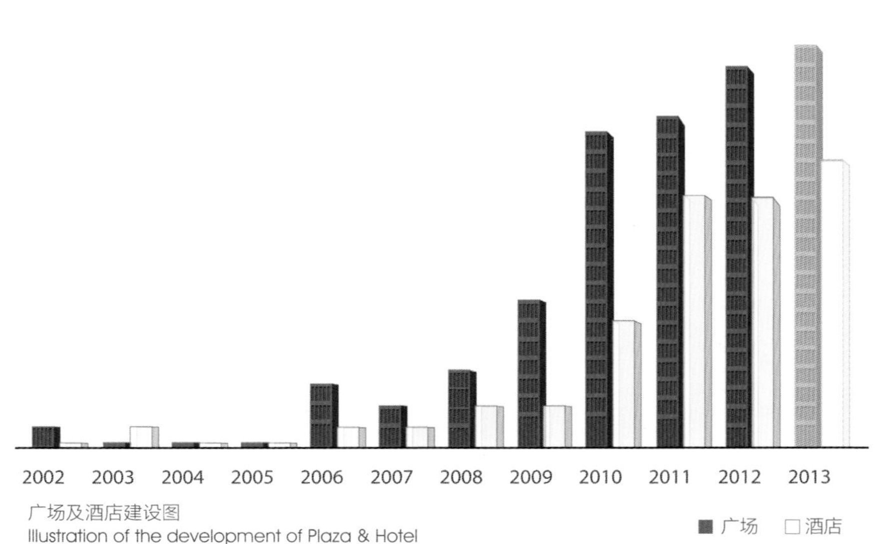

广场及酒店建设图
Illustration of the development of Plaza & Hotel

In 2009, Wanda's commercial real estate business advanced in spite of unfavorable environment and intensified its development. 8 Wanda Plazas were opened in 2009. In 2010, the following year, Wanda's development and construction was further accelerated. In that year, 15 Wanda Plazas and five five-star hotels were opened. That is to say, the numbers of both commercial buildings and hotels were nearly doubled at the same time. In 2011, 14 Wanda Plazas and 12 hotels were opened. Namely, the number of hotels opened for business was more than doubled again! After three consecutive years of accelerated development and construction since 2009, 18 Wanda Plazas and 12 five-star hotels were opened in 2012. Thereafter, the growth rate of Wanda's own properties will be maintained at a steady level, with 17 to 20 Wanda Plazas and 10 to 12 five-star hotels to be opened each year.

万达2009年的加速发展，对万达商业地产的人才专业结构，提出了更高的要求。万达商业规划研究院在成立两年的经验基础上，2009年完成了人员初步配置，形成了规划、建筑、结构、机电、内装、景观全专业的人员配备，人员编制为80人；2010年，增加夜景照明专业，成立了节能研究所、旅游规划所，编制扩大到125人；2011年，随着万达在文化旅游领域的投入和发展，万达商业规划研究院成立了文化旅游分院，全院编制为225人；2012年，伴随万达文化旅游项目的全面展开，根据对秀场、电影乐园、主题公园、滑雪场等特种类专业的需求，全院再次扩编，发展到300多人。万达商业规划研究院负责集团商业、酒店及文化旅游全部持有物业的规划设计管控。

Wanda's accelerated growth in 2009 posed a higher requirement to the specialties structure of Wanda's talents in commercial real estate. With the experience accumulated in the two years after its establishment, Wanda Commercial Planning & Research Institute finished its initial personnel allocation in 2009. It had a complete personnel structure covering all specialties including planning, architecture, structure, MEP, interior decoration and landscaping, with a headcount of 80. In 2010, nightscape lighting was added as a new specialty and the Department of Energy Conservation Research and the Cultural Tourism Department were established, with the headcount expanded to 125. In 2011, with Wanda's input and development in the cultural tourism field, the Institute established its Cultural Tourism Branch, with the headcount of the entire Institute reaching 225. The year of 2012 saw the all-round development of Wanda's cultural tourism projects and thus there emerged the demand for special types of specialties, including the show theatre, movie amusement park, theme park and ski resort. Therefore, the headcount of the Institute was further enlarged to over 300. The Institute is now responsible for the planning, designing and controlling of all of Wanda's own properties including commercial, hotel and cultural tourism projects.

万达2009年的加速发展，对万达商业地产开发项目的整体管控，以及对保证项目建设速度及项目建设确保设计品质与建造质量，提出了前所未有的要求。2009年也是万达商业规划院标准化建设、信息化建设系统展开的元年。2009年万达商业规划研究院为应对集团开发项目数量的快速增加，执行团队新员工数量的高速膨胀，由万达商业规划研究院组织编写了《万达商业规划管理标准》和《万达商业规划技术标准》，成为集团制度的一部分，同时也为今后大规模的万达广场及五星级酒店的设计及设计管控打下了坚实的基础，为2010年集团制度的修编提前做好了准备。

Wanda's accelerated growth in 2009 posed unprecedented requirements to the overall control of Wanda's commercial real estate development projects and the quality of designing and construction at the same time of guaranteeing the speed of project construction. 2009 is also the first year for Wanda Commercial Planning & Research Institute to systematically develop its standardization and informationization. With the number of projects developed by the Group enlarged rapidly and the number of new employees in the project teams increased sharply, the Institute organized the compilation of the *Management Standards for Wanda Commercial Planning* and the *Technical Standards for Wanda Commercial Planning*, which also became a part of the systems of Wanda Group as a whole. This also laid a solid foundation for the designing and design controlling of a large number of Wanda Plazas and five-star hotels in future and paved the path for the revision of the Group's systems in 2010.

光阴似箭，过去的2009年在万达商业规划研究院的发展中具有特殊的意义。为此，从2009年开始，我们在本次出版的年鉴图册中，特别收录了每一年万达商业规划研究院所有同事的照片及正式员工姓名。同时借此图书出版之际，向那些曾经一同为万达商业规划研究院工作，为万达的事业辛勤付出的规划院的同事们表示诚挚的敬意，并对所有为万达工作过的同仁及关注和支持万达的朋友们表示由衷的感谢！

万达集团董事长王健林指导规划院工作
Wang Jianlin guiding Wanda Commercial Planning & Research Institute

Time flies fast. The bygone year of 2009 is of special significance to the development of Wanda Commercial Planning & Research Institute. Therefore, since 2009, we specially collected the photos of all colleagues in the Institute and the names of all full-time employees each year in the picture part of this yearbook. At the same time, on the occasion of the publishing of this book, we hereby extend our sincere respect to those who worked together with us for the Institute and gave diligent contribution to Wanda's cause and also express our heartfelt gratitude to all our ex-colleagues and those friends who gave their attention and support to Wanda!

大连万达商业地产股份有限公司高级总裁助理
万达商业规划研究院院长
赖建燕
Lai Jianyan
Senior Assistant of the President of Dalian Wanda Commercial Estate Co., LTD
President of Wanda Commercial Planning & Research Institute

万达商业综合体构成概述
THE COMPOSITION OF WANDA COMMERCIAL COMPLEX

万达商业综合体是万达集团独创的新一代商业地产模式化产品，经多年的持续发展并逐步完善。它是将城市活动中多种不同功能空间进行有机地组合，并在不同功能间建立一种空间依存、价值互补的能动关系。

Wanda commercial complex is a new modeled product of commercial property independently invented by Wanda Group and has been evolving for years. It organically integrates the spaces with different functions and creates between them a dynamic relation featuring spatial interdependence and mutual complement in value.

万达商业综合体是城市综合体的一种，是以万达购物中心为主，城市商业街为辅，同时设置高星级酒店、甲级写字楼、高级公寓及居住配套的城市综合体（图1）。

As an urban complex, Wanda commercial complex centers on Wanda shopping center, which is flanked by urban commercial streets and boasts luxury hotels, class A office buildings, upscale apartment buildings and residential supporting facilities (Figure 1).

以万达广场命名的万达购物中心是万达商业综合体中的商业核心。

Wanda Shopping Center, named after Wanda Plaza, is the commercial core of Wanda commercial complex.

图 1：万达商业综合体业态构成图
Figure 1 Illustration of the Composition of Wanda Commercial Complex

1 万达商业综合体的主要特征

1 Pronounced Features of Wanda Commercial Complex

1.1 功能复合性

1.1 Multiple Functions

多功能复合开发是万达商业综合体的基本特征，综合体涵盖商业、商务、办公、居住、餐饮、文化娱乐等各种功能，满足消费者集购物、休闲、娱乐于一体的“一站式消费”需求。

The development of multiple functions is considered as the essential feature of Wanda commercial complex, making the complex a venue for manifold activities including business, commerce, office, residence, dining and cultural experience to the needs of consumers for "one-stop consumption" comprised of shopping, leisure and recreation.

1.2 空间集约化

1.2 Optimal Space Utilization

随着我国城市化进程的发展，城市土地资源紧缺和人口密度高的矛盾不断加剧，提高土地利用的集约度成为城市发展的必然选择。因此我国绿色建筑的首要标准就是节地，万达商业综合体的建设发展符合国策，以高度的空间集约化实现城市空间的高效利用，主要体现在建筑密度大、容积率高。

The ongoing urbanization in China aggravates the contradiction between the shortage of land resources and high population density in urban areas, thus making the optimization of land utilization indispensable in city development. For this reason, land conservation becomes the prime norm of developing green building in China. Wanda commercial complex has been built and developed in line with this national policy. The compactness and high capacity rate of Wanda commercial complex show how we have efficiently used urban space.

1.3 注重公共性

1.3 Emphasis on Public Property

万达商业综合体具有城市功能属性，是向城市开放的。万达商业综合体设计注重城市公共空间的规划设计，其外部空间是同城市空间相互融合的。每一座万达广场的主要出入口均设有大型的城市商业广场；高级酒店规划有前广场、后庭院；步行商业街营造出更为开放、多样性的休闲、购物空间，由此形成城市中的不同特色的多个节点空间，极大地增加了城市生活的丰富性。

Performing parts of urban functions, Wanda commercial complex is open to the whole city. Hence, we place great emphasis on the planning of urban public space in the design of Wanda commercial complex because its external space forms a part of the city. Large-scale urban commercial squares are set up to lead the way to main entrances and exits of all Wanda Plazas; luxury hotels are attached with front squares and backyards, and pedestrian commercial streets create more open and diverse space for recreation and shopping. In this way, a space with multiple nodes and flavors comes into being inside the city, immensely spicing up urban life.

1.4 城市中心

1.4 City Center

万达商业综合体作为城市生活集中的载体，其多样化的功能和庞大的建筑群体规模以及便利的城市交通形成集聚效应。一座万达广场就能形成一处城市中心。

As a central place of urban life, the Wanda commercial complex with multiple functions also enjoys enormous scale of building clusters and convenient city traffic, which eventually generate an agglomeration effect, making every Wanda Plaza a city center.

2 万达商业综合体的发展

2 The Development of Wanda Commercial Complex

万达商业综合体发展到现在，就其产品形式而言共经历了三代：2000～2003年为第一代单店模式，建筑表现为单体商业楼，总面积5万平方米左右，业态构成为主力店加精品店的模式。2003～2004年，万达商业综合体开始进入短暂过渡的“组合店”模式，每个项目有四、五个独立商业楼，通过室外步行街连起来做成一个商业广场，总规模约15万平方米左右。2004年后至今是第三代大型城市综合体的产品模式。第一、二代万达商业综合体主要依托于城市中心地带，借助城市传统商圈的力量来发展，但第三代万达商业综合体已经能够依靠自身的商业模式和规模效应独立创造出城市新中心和新商圈。

Three generations of Wanda commercial complex have evolved over the years. The 1st generation from 2000 to 2003 was Single Building Model. It featured a single commercial building covering a total area of about 50,000 m^2. Types of business included anchor stores and boutiques. 2003 and 2004 witnessed the development of the 2nd generation,

Combined Buildings Model, which was transient and transitional. Each project was composed of 4 or 5 separate commercial buildings connected by an outdoor pedestrian street to form a commercial plaza, spanning over a total area of about 150,000m^2. The 3rd generation, Urban Complex, has been developing since 2004. The 1st and the 2nd generations of Wanda commercial complex were mainly reliant on the urban center area, seeking growth by dint of the traditional urban commercial districts. In contrast, the 3rd generation Wanda commercial complex is capable of nursing new city centers and new commercial districts by virtue of its own business model and scale effect.

从2004年，自宁波鄞州万达广场开始，万达商业地产进入到第三代产品—城市综合体，从项目的选址到开发规模、产品形态都发生了较大的变化。第三代万达商业综合体在国内首次创新引入了室内步行街的概念。购物中心突出的特点是一街带多楼的空间组合形态。一街即为室内步行街，其贯穿整个建筑，通常为3～4层，室内步行街沿街一条动线保证了所有店铺和主力店人流均可到达。多楼指的是百货楼、娱乐楼、综合楼等，通常为5～6层。

The 3rd generation property model—Urban Complex began with Ningbo Yinzhou Wanda Plaza in 2004. Everything, from site selection to development scale and the form of product, has undergone significant changes. The 3rd generation Wanda commercial complex first introduced the concept of indoor pedestrian street to China. The hallmark of the shopping center is the spatial configuration of "one street promoting numerous buildings". Usually 3 to 4 stories high, the "one street", i.e.. indoor pedestrian street, runs through the entire complex and forms a circulation, ensuring easy accessibility of all stores and anchor store buildings. "Numerous buildings" indicate shopping departments, recreation buildings and comprehensive buildings, which are usually 5 to 6 stories high.

正在研发中的第四代万达商业综合体是在第三代万达商业综合体基础上优化衍生出的新型商业——文化旅游城。新式文化旅游城秉承第三代万达商业综合体以室内街为灵魂的规划设计精髓，以一条商业步行街为纽带，分别串联起室内主题公园、电影乐园、室内水上乐园、儿童科技乐园、剧场等不同的业态，创造性地发展了更多更丰富的空间组合形态，为商业与旅游的相互结合、促进找到了契合点。第四代万达的创新产品从国际化的发展角度必将引领国内的商业、旅游发展新模式。

The 4th generation Wanda commercial complex under research is a new type of business: culture and tourism city. It incorporates the essence of the planning and design of the 3rd generation Wanda commercial complex, the soul of which is indoor pedestrian street. Acting as a bond, the commercial pedestrian street links up different business types, including indoor theme parks, movie parks, indoor water parks, kids' science and technology parks and theaters. More and richer forms of spatial configuration are innovated, enhancing the integration and the mutual promotion of business and tourism. The innovative 4th generation Wanda commercial complex will surely be the arrow head for developing fresh business and tourism models in China from the perspective of globalization.

3 万达商业综合体的构成
3 The Composition of Wanda Commercial Complex

3.1建筑构成
3.1 The Composition of Architectures

万达商业综合体通常由购物中心、高星级酒店、甲级写字楼、高级公寓、城市商业街、住宅等建筑构成。

Wanda commercial complex is generally made up of shopping centers, luxury hotels, class A office buildings, upscale apartment buildings, urban commercial streets, residential buildings.

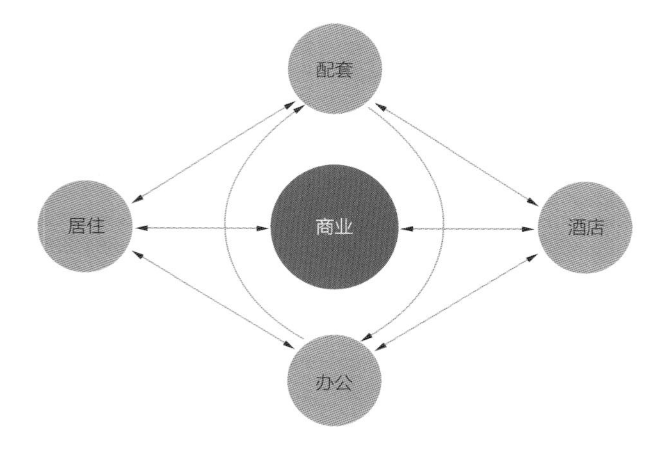

图 2：万达商业综合体业态关系示意图
Figure 2 Illustrations of the Relations between Different Businesses in Wanda Commercial Complex

3.2 业态构成

3.2 The Composition of Business Types

万达商业综合体业态包含商业、办公、酒店、居住及其他配套，各业态之间形成有机联系、相互依托、互为能动（图2）。万达购物中心的业态由主力店、次主力店和室内步行街商铺构成。主力店业态通常为万达百货、超市、电器、万达影城、大歌星KTV、大玩家超乐场、酒楼。次主力店为国际知名时尚品牌，吸引大量年轻时尚消费群体。室内步行街通常为3层，一、二层主要为精品零售商铺，三层为餐饮。室内步行街是购物中心内最活跃、最重要的业态形式。

In Wanda commercial complex there are varied business types such as commerce, office, hotel, residence and other supporting facilities, which are interconnected, interdependent and interactive (Figure 2). Businesses in Wanda shopping center are comprised of anchor store buildings, sub-anchor store buildings and an indoor pedestrian street. The anchor store buildings contain Wanda Dept. Store, supermarkets, electronic appliance buildings, Wanda Cinemas, Super Star KTV, Super Player Park and restaurants. World-famous fast fashion brands entice legions of young and stylish consumers. The indoor pedestrian street normally has 3 stories with boutique retail stores occupying the first and second floors and restaurants occupying the third floor. The indoor pedestrian street is the most active and important business type in Wanda Plaza.

万达商业地产经过长期的发展摸索，依靠第三代万达商业综合体这种独创的成熟产品模式的开发建设实现了万达商业地产自身快速发展，而且还在带动项目所在城市的产业调整、拉动消费需求、完善城市区域功能、创造就业岗位和税源等四个方面实现了巨大的社会效益。目前在全国已经建成的60余个万达广场，建成即开业、开业即满铺，为城市功能的完善和城市整体发展做出了巨大贡献。

After many years of growth and exploration, Wanda Commercial Properties Co., Ltd. has achieved rapid development by creating and developing a mature property model, the 3rd generation Wanda commercial complex. In addition, the company has brought about considerable social benefits to the cities by pressing forward the industrial restructuring, driving the consumption demand, improving the regional functions and creating job opportunities and tax revenue. Currently, there are over 60 Wanda Plazas across the country, all of which were put into operation upon completion and became packed upon opening, making great contribution to the improvement of urban functions and the development of the entire city.

万达商业规划研究院副院长

黄大卫

Huang Dawei

Vice President of Wanda Commercial Planning & Research Institute

万达商业综合体总图规划设计
THE GENERAL LAYOUT PLANNING AND DESIGN OF WANDA COMMERCIAL COMPLEX

商业综合体总图规划设计，与其他类型的建筑总图规划截然不同。商业建筑一定是“业态规划在先，建筑规划在后”。

The general layout planning and design of commercial buildings, which is distinct from that of other buildings, must abide by the principle of "business type planning coming before construction planning".

总图首先是业态规划，这是万达商业地产总图规划设计的基本原则，也是万达在具体实施并成功运营万达广场多年的经验总结，更是万达在商业地产领域立于不败之地的不二法宝。业态规划是在万达的“订单地产”强大保证前提下，做出的前瞻性规划。它把酒店、商务、商业、娱乐、休闲、居住等各个业态高效合理地组织起来，预估了从前期拿地、建设到后期的管理、销售的全过程均好性，是综合了经济效益与社会效益的顶层设计。

Business type planning must be prioritized in the layout planning and design. This is not only a fundamental principle upheld by Wanda Group in developing commercial properties, but also a valuable conclusion reached after many years of successful operation of Wanda Plaza. More importantly, it is our silver bullet that has made us invincible in the field of commercial properties. Taken as a forward-looking planning based on the strong backing of the "order-based properties" of Wanda Group, business type planning efficiently synthesizes different businesses like hotel, commerce, business, recreation, leisure and residence and aims to achieve a better sharing of resources over the whole process from acquiring land, construction to management. It is a top-level design integrating both economic and social benefits.

与之相反，很多房地产企业转过来做商业中心的时候，对规划设计是不重视的，或重视的是一种外形，或只看重建筑师的名望。他会告诉你，我的商业综合体是请一个世界著名的设计师设计的，这个设计师在亚洲或者是在欧洲很有名。这样是错的，建筑师是很难做好商业综合体的规划设计的。

In contrast, when it comes to developing commercial centers, many real estate companies either think little of planning and design or only emphasize external appearance or the fame of architects. They take comfort in the idea that their commercial complex is designed by a world-renowned planning institute or an architect well known in Asia or Europe. But they are wrong because the chances are slim that any architect could do a good job of commercial complex planning and design.

万达商业综合体总图规划设计的原则是：交通优先，功能分区，突出商业，公共空间。

The layout planning and design of Wanda commercial complex follows the principle of traffic prioritization, function zoning, business prominence and public space.

交通优先对于万达的城市综合体有三个层次的含义：一是综合体区域与区域外的城市级交通联系；二是基地内部的交通组织；三是购物中心内部动线的合理组织。只有三个层次都合理解决才是真正意义的交通优先。

Traffic prioritization indicates three meanings for Wanda urban complex. First, it means the traffic connections between the urban complex and the city; second, it means the traffic within the complex; and third, it means a properly designed circulation inside the shopping center. Only when all of the three are properly addressed can we truly realize the principle of traffic prioritization.

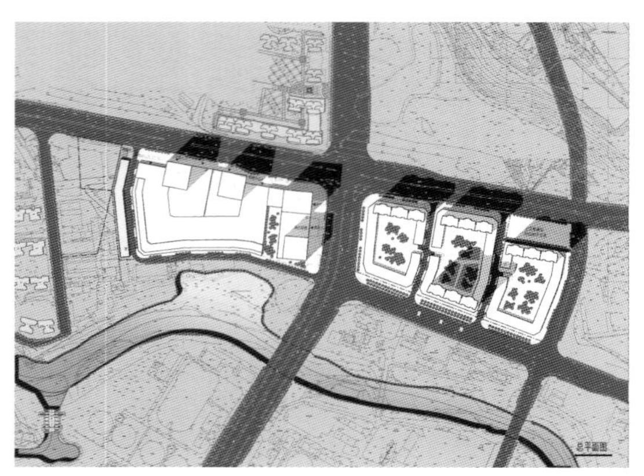

南宁万达广场总平面图
The master plan of Nanning Wanda Plaza

功能分区是业态规划的重要体现，是指从城市综合体总体而言，按照功能在总体规划中进行分区，并在规划中体现综合体各功能区域与城市在交通、景观、规划体系、文化等各方面的关系。功能分区的原则第一是要按照综合体中业态彼此之间的功能对应关系布置；第二是按照房地产开发的资本运作要求布置和取舍各功能业态；第三是按照建筑规划设计的基本规律进行设计。

Function zoning is a significant embodiment of business type planning. The zoning is made according to the functions under the overall planning in terms of the entire urban complex. At the same time, the planning must give expression to the relations between the different functions of the urban complex and the city regarding traffic, landscape, planning system and culture. With respect to function zoning, there are three principles to be followed. The first is to deploy business types according to their corresponding relations in the complex; the second is to arrange and choose businesses with different functions as required by the capital operation of property development; and the third is to design in light of the basic rules of construction planning and design.

万达商业综合体是以商业为主导，涵盖了商业、商务和居住三大房地产开发板块的城市综合体。商务与居住虽然都是城市综合体的组成部分，而且都可以独立存在，但是任何居住区和商务办公区都需要商业服务设施，且与商业属于互相依存的形式。在大型社区中，大型商业中心及中小型商业服务配套满足居住、商务的服务需求，是必不可少的重要部分。同时，居住和商务业态无论从规划选址、建造交付，到入住管理都相对简单，而商业建筑对于交通、选址、招商、建造乃至后期经营管理都相对复杂很多。而商业项目的兴旺，可以带动周边居住及商务区的发展和生活品质的提升。在全国很多万达广场周边的楼盘，都打出“距某某万达广场X公里”的广告；同时，很多万达广场从拿地建设到开业，都可以大幅提高周边物业价值。

Wanda commercial complex covers three major types of real estate development, namely, commerce, business and residence, among which commerce is predominant. Although all of the three are also parts of the urban complex and can stand alone, residential and business office areas need to be supplied with commercial service facilities and consequently are interdependent with the commerce areas. Extensive communities cannot be dispensed with big commercial centers as well as small and medium sized commercial service supporting to meet the needs for residential and business services. Every step of residential and business property development, no matter it is planning, site selection, construction, turning over, occupancy or management, comes easier in comparison with the development of commercial buildings, which seems much more complicated in all aspects including traffic, site selection, attracting investment, construction and operation. Furthermore, thriving commercial projects can boost the development of the surrounding residential and business districts and promote the quality of life. For example, billboards with words like "X kilometers away from X Wanda Plaza" have been put up in the properties being developed in the vicinity of Wanda Plaza. Besides, the property value of the areas nearby Wanda Plazas has been driven up substantially during the time when Wanda Plaza was launched and put into operation.

万达集团的业态规划以商业为主导，商业建筑是大型公共建筑，同时也是人员密集场所，需要有集中的大型公共空间满足商业功能、人员集会等需要。这些室内外公共空间涵盖：商业广场、集中绿地和景观水系、室内中庭、室内连廊等，其中商业广场是配合大型商业建筑的人员集散和室外公共商业活动的主要场所，集中绿地、景观水系等是街区性商业、餐饮酒吧类商业的室外辅助空间，而室内公共空间更是商业密不可分的组成部分。万达对于公共空间的重视是万达商业综合体成功的重要因素。

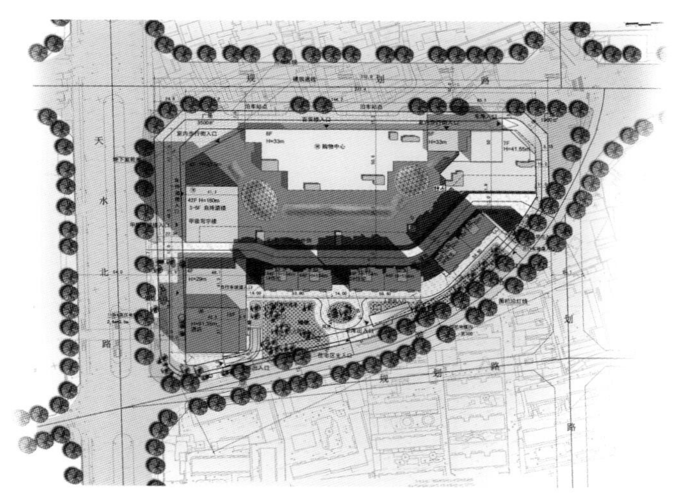

兰州万达广场总平面图
The master plan of Lanzhou Wanda Plaza

The business type planning of Wanda Group is dominated by commerce. Commercial buildings are grand public buildings with heavy flows of people. Therefore, concentrated commodious public spaces are needed to satisfy the needs for business and gatherings. These outdoor and indoor public spaces include: commercial squares, indoor atrium with concentrated green spaces and landscape water systems and indoor vestibules. Commercial squares, in support of large commercial buildings, function as important venues for gatherings and outdoor public commercial activities. Concentrated green space and landscape water systems are auxiliary spaces of neighborhood commerce and businesses like restaurants and pubs. Indoor public spaces make an integral part of commerce. The great attention paid by Wanda Group to public spaces greatly contributes to the success of Wanda commercial complex.

万达的总图规划设计都是由万达商业规划研究院来完成的，这是保证万达商业综合体品质与速度的前提。同时以万达商业规划研究院多年的经验，除了上述的原则，进行商业综合体总图规划设计时还应注意城市选址、市政条件、规划条件以及同质竞争等问题。总图规划是整个商业综合体项目的第一次主动创造，是后续的单体设计的框架，是保证商业综合体项目成功迈出的最重要的一步。

The responsibility of the general layout planning and design of Wanda Group is shouldered by Wanda Commercial Planning & Research Institute, a prerequisite to guarantee the quality and speed of Wanda commercial complex development. In view of the experience collected by Wanda Commercial Planning & Research Institute over the past years, other matters except for the above-mentioned principles, such as city selection, municipal administration conditions, planning conditions and homogeneous competition, also need to be given proper consideration in the general layout planning and design of a commercial complex. The general layout planning, as the first initiative creation of the whole commercial complex project, offers the framework for the subsequent single building designs and counts as a crucial step forward in securing successful commercial complex projects.

万达商业规划研究院副院长
朱其玮
Zhu Qiwei
Vice President of Wanda Commercial Planning & Research Institute

PART 1 万达广场 WANDA PLAZAS

南京建邺
万达广场
NANJING JIANYE WANDA PLAZA

天幕夜景

室内步行街

南京建邺万达广场位于南京市建邺区，总用地面积 27 万平方米，总建筑面积 97 万平方米，包括 27.5 万平方米购物中心、5 万平方米超五星级酒店、34.5 万平方米写字楼、30 万平方米高档公寓，是南京市最大的商业项目。

Nanjing Jianye Wanda Plaza is located in Nanjing City, Jianye District, with a total land area of 270,000 m^2, a total construction area of 970,000 m^2, including 275,000 m^2 shopping center, 50,000 m^2 of super five-star Hotel, 345,000 m^2 office building, 300,000 m^2 of high-grade apartments. It is the largest commercial project in Nanjing City.

购物中心于 2009 年 12 月 18 日盛大开业，由室内步行街、百货楼、娱乐楼、大型超市以及配套的地下停车场、地下设备用房、管理用房等共同构成，包括万达百货、万达影城、大歌星 KTV、大玩家超乐场、国美电器、超市等多种业态。

The shopping center opened in Dec. 7, 2009. The shopping center consist of indoor pedestrian street, department store, supermarket and entertainment buildings, related underground parking garage, underground equipment rooms, management offices, including Wanda Dept. Store, Wanda Cinemas, Super Star KTV, Super Player Park, Gome, supermarket and other industrial businesses.

超五星级南京万达希尔顿酒店于 2011 年 11 月 16 日正式开业，拥有 360 间奢华客房，并配有高规格的餐饮、宴会、康体设施。

Super five-star Nanjing Wanda Hilton Hotel was officially opened in November 16, 2011, which has 360 luxury rooms, and is equipped with a high variety of dining, banquet, sports facilities.

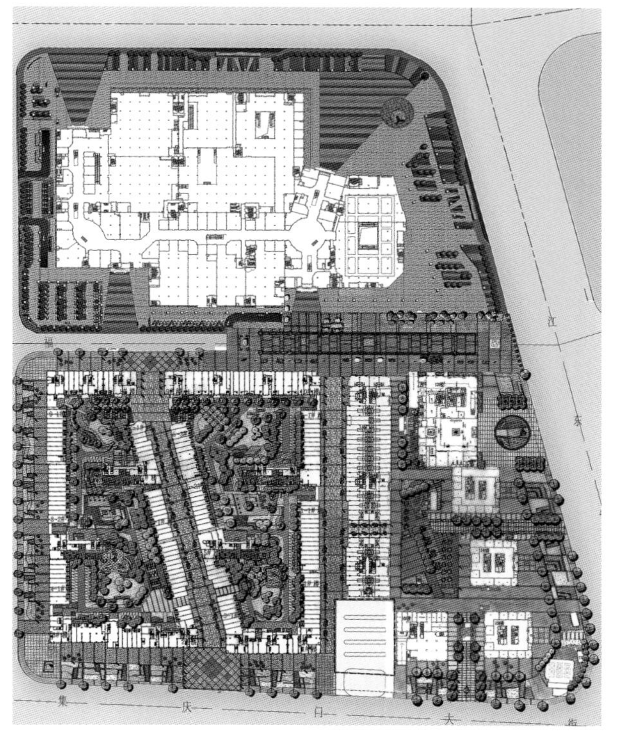

总平面图

入口夜景

室内步行街

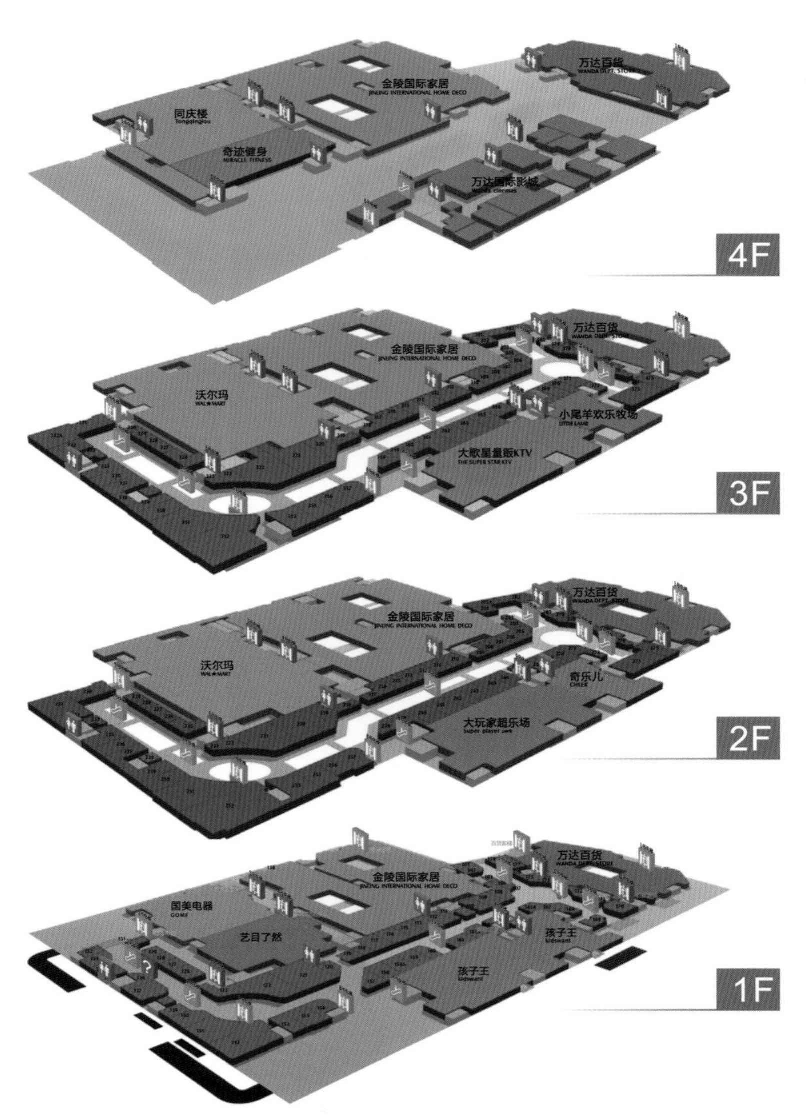

品牌落位图

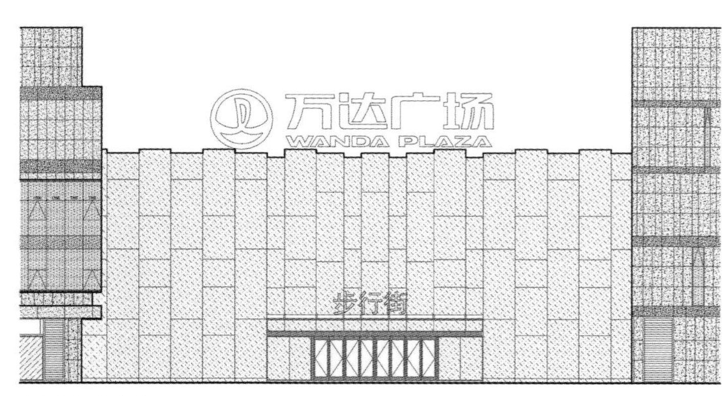

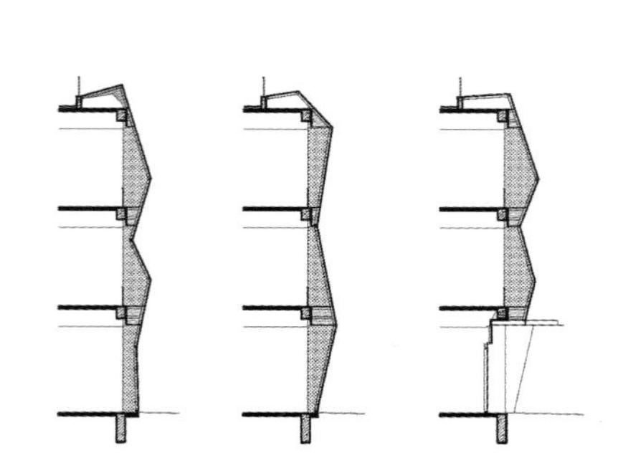
步行街门头立剖面图

广场外立面

大歌星 KTV

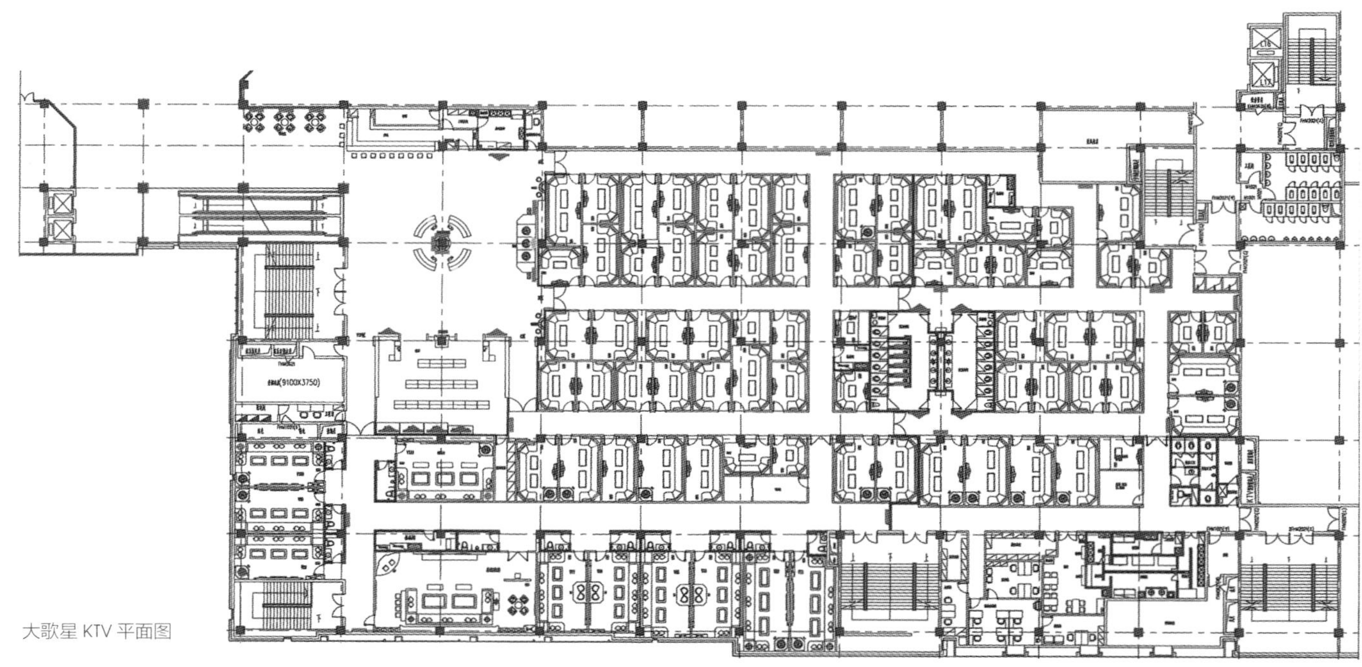

大歌星 KTV 平面图

青岛CBD 万达广场

QINGDAO CBD WANDA PLAZA

广场日景

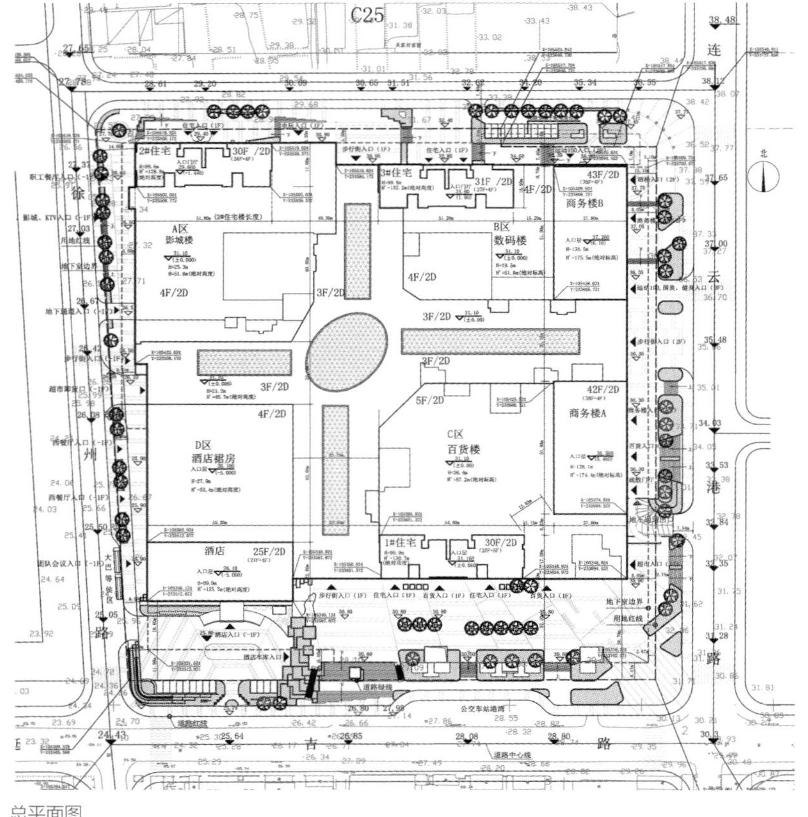

总平面图

室内步行街

Qingdao CBD Wanda Plaza has been developed as a 3rd generation Wanda product accommodating large scale shopping mall, five-star hotel, office buildings and apartments, and interior shopping street. Simple and modern approach in facade design, geometric composition and details in the turning of horizontal stripes at the corner of the office tower gives a strong character to this shopping facility. Qingdao CBD Wanda Plaza has developed as a district center, Including shopping, business and residential.

青岛 CBD 万达广场坐落于青岛市市北区中央商务区（CBD）的核心地段，西起山东路，东至福州路，南起延吉路，北至辽源路，紧邻市北区政府和青岛市图书馆。项目总建筑面积 38 万平方米，商业面积 20 万平方米，集合了大型购物中心、五星级酒店、写字楼、住宅楼、室内步行街等多种业态，为万达第三代城市综合体建筑。青岛 CBD 万达广场业态品牌结构完整、品类丰富，以餐饮经营为主，并结合了娱乐休闲类、生活配套类商铺及品牌，主力店则包含了万达百货、万达影城、艾美酒店、超市、国美电器、大歌星ＫＴＶ、大玩家超乐场等国内外知名连锁商家。整个项目外立面大气，裙房部分体块穿插组合有序，高层塔楼的横线条在转角部位的处理别具特色。该项目形成集购物、商务、居住为一体的区域性中心。

Qingdao CBD Wanda Plaza is located in the CBD area of North District, Qingdao City. The site of the project is adjacent to the municipal government of the north district, defined by the Shandong road the west, Fuzhou road to the east, Yanji road to the south and the Liaoyuan road to the north. With total floor area of 380,000 m^2, of which 200,000 m^2 of commercial section,

GIORDANO
KFC
Coodoo
小淑女与约翰

室内步行街中庭

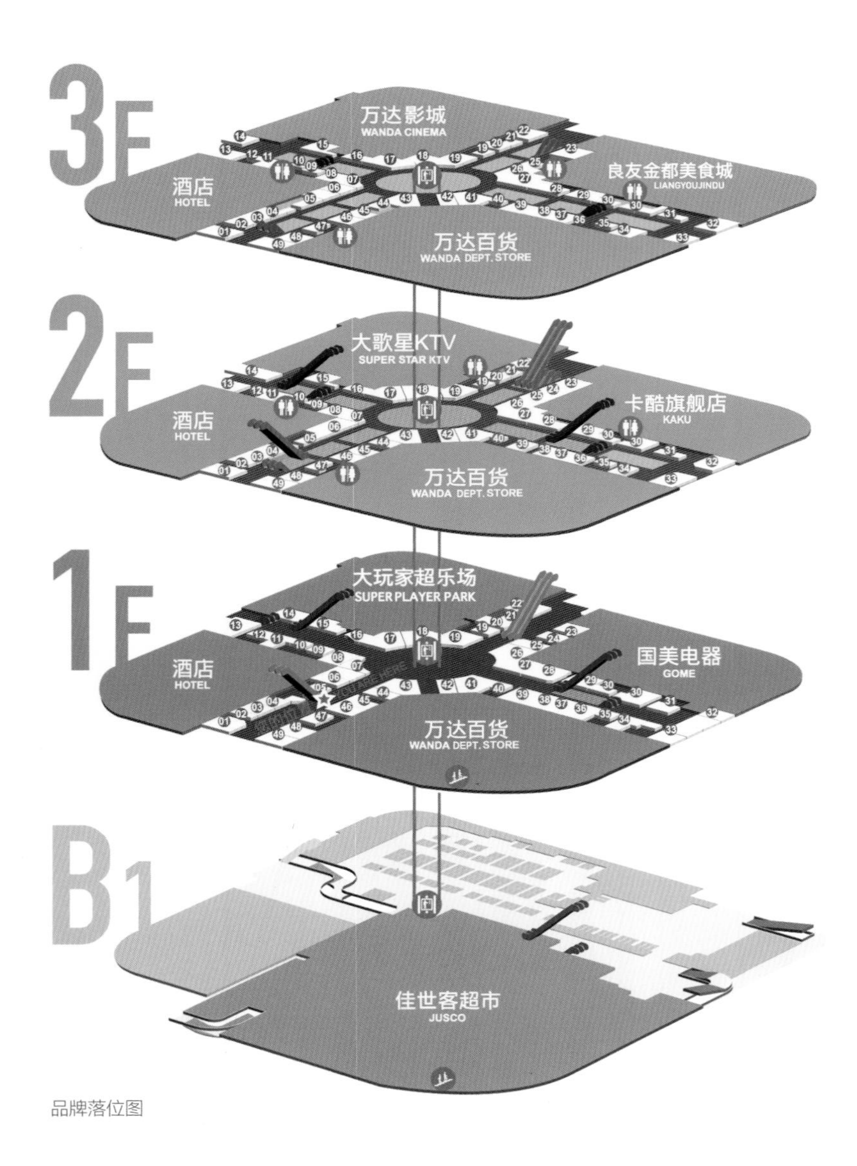
3F
万达影城
WANDA CINEMA
酒店
HOTEL
良友金都美食城
LIANGYOUJINDU
万达百货
WANDA DEPT. STORE
2F
大歌星KTV
SUPER STAR KTV
酒店
HOTEL
卡酷旗舰店
KAKU
万达百货
WANDA DEPT. STORE
1F
大玩家超乐场
SUPER PLAYER PARK
酒店
HOTEL
国美电器
GOME
万达百货
WANDA DEPT. STORE
B1
佳世客超市
JUSCO

品牌落位图

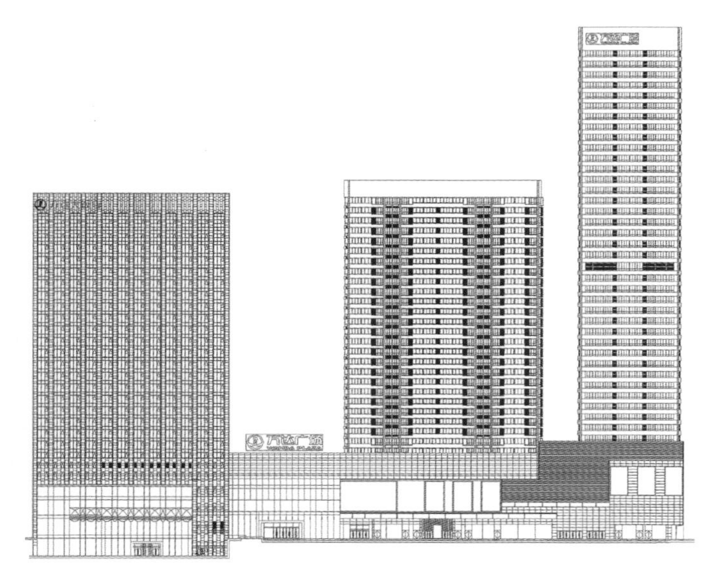

广场立面图

LE MERIDIEN
青岛万达艾美酒店

广场夜景鸟瞰

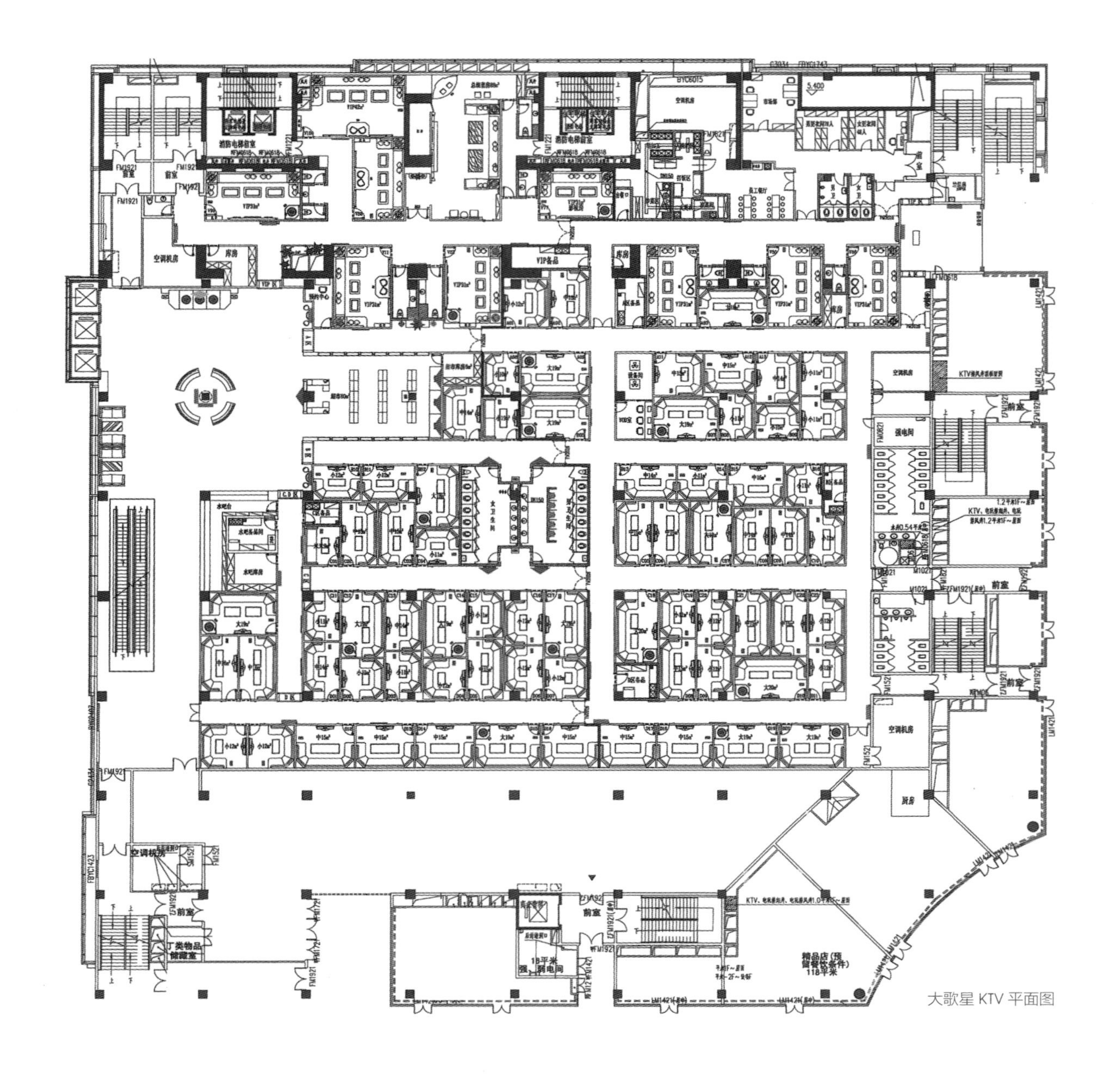

大歌星 KTV 平面图

步行街中庭

大歌星 KTV

上海周浦万达广场

SHANGHAI ZHOUPU WANDA PLAZA

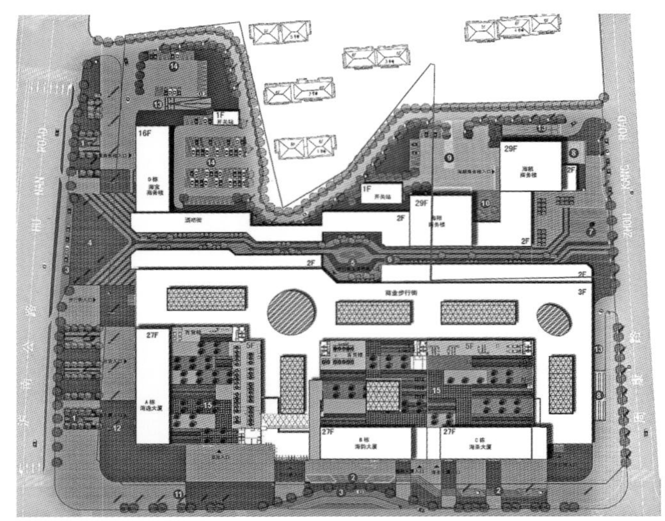

总平面图

上海周浦万达广场位于上海市南汇区周浦镇，基地南至年家浜路，西邻沪南公路，东至周康路，地块面积 6.37 公顷，用地规划性质为商业金融办公用地。容积率：4.0，建筑密度：54%。

Shanghai Zhoupu Wanda Plaza is located in Zhoupu Town Nanhui District. It lies south to NianJiaBang road, west to Hunan road, east to Zhoukang road, with a land area of 637,000 m^2. The land's plan character is for commercial, financial and office. Volume rate: 4.0, building density: 54%.

上海周浦万达广场由百货楼、时尚楼、娱乐楼和商业步行街、酒吧街及高层塔楼等组成，包括超市、万达百货、大玩家超乐场、万达影城、健身、电器、大歌星 KTV、运动用品等十大主力店，超市设置在地下一层。

Shanghai Zhoupu Wanda Plaza contains department stores, fashion houses, fun houses, commercial pedestrian streets, bar streets, top floor sets and other components, which including supermarket, Wanda Dept. Store, Wanda Cinemas, Super Player Park, fitness centre, appliance store, Super Star KTV, sports supplies store as ten main stores. The supermarket were set up in the basement.

该项目是万达集团在上海投资的第二个城市综合体，2009 年 9 月开业，至今仍为该地段的商业中心。

The project is the second city complex invested in Shanghai by Wanda Group, opened in September, 2009, which has become the business center of this area.

广场入口

60
周浦万达广场
9.19盛装开业
GRAND OPENING
TESCO 乐购
JACK & JONES
JACK WALK
ONLY

40

室内步行街采光顶

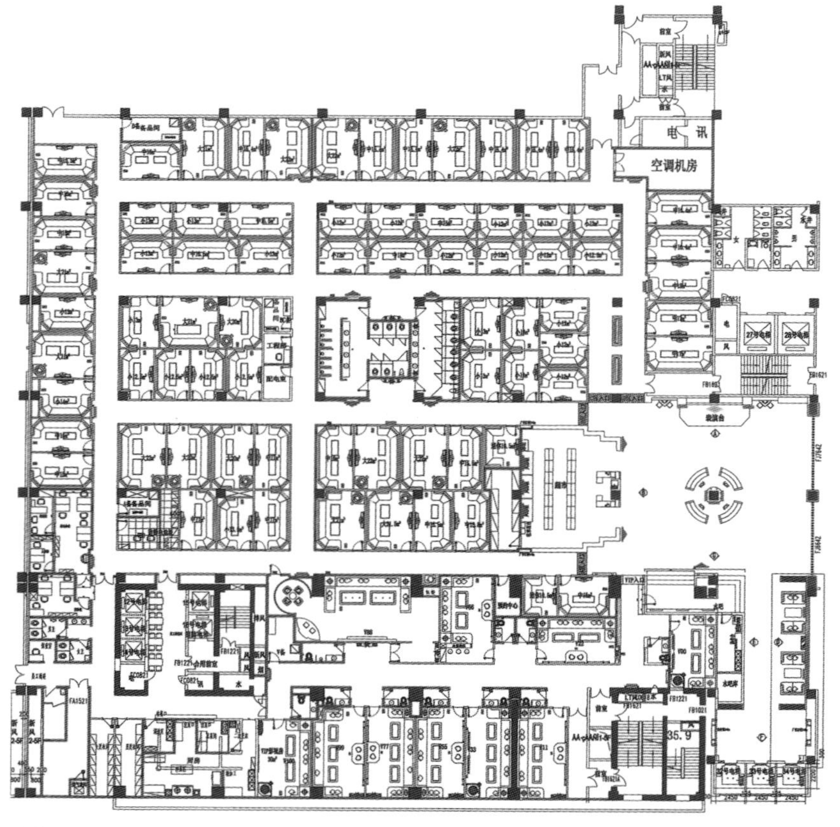

大歌星 KTV 平面图

广场日景

室内步行街

室内步行街

苏州平江
万达广场
SUZHOU PINGJIANG WANDA PLAZA

苏州平江万达广场位于苏州市平江新城规划商务核心区域，用地西邻人民路，南临 312 国道，北侧为规划道路，东侧为河道及规划绿地。项目总占地面积为 6.4 万平方米，用地南北长约 400 米，东西宽约 214 米，南侧为 15 米宽城市绿化带，自然景观优势良好。

Suzhou Pingjiang Wanda Plaza is located in the city of Suzhou Pingjiang new town planning business core area. The land is adjacent to the east of Renming road, south to the state road No. 312, north to the planning road, east to the river and planning green spaces. The project boundray area is 64,000 square meters, with 400 m length from north to south, with a east-west width of approximately 214 m. and in the south side, it owns 15-meter wide city green belt. The natural landscape is fantastic.

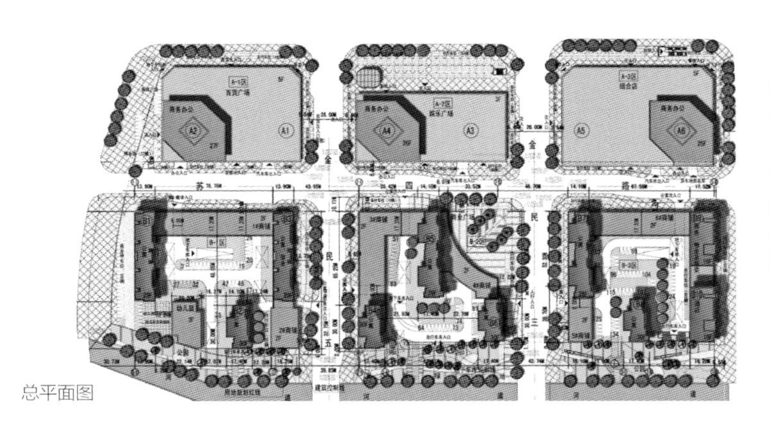

总平面图

本项目 2009 年 9 月盛大开业，项目总建筑面积约 45 万平方米；其中地上建筑面积 35.0 万平方米，地下建筑面积 9.7 万平方米，分为大商业和公寓、酒店、商务办公、商铺。建筑高度 100 米；其中地上 27 层，地下 2 层。

This project grandly opened in September, 2009. The project's total construction area is about 450,000 m^2; with ground area of 350,000 m^2, area of 97,000 m^2 underground. It is divided into commercial, apartments, hotels, offices, shops. The building's heighth is 100 meters; with the ground 27 layers, underground 2 layers.

广场日景

广场夜景

万千百货
VAN'S DEPT. STORE
VERO MODA
万达广场

广场夜景

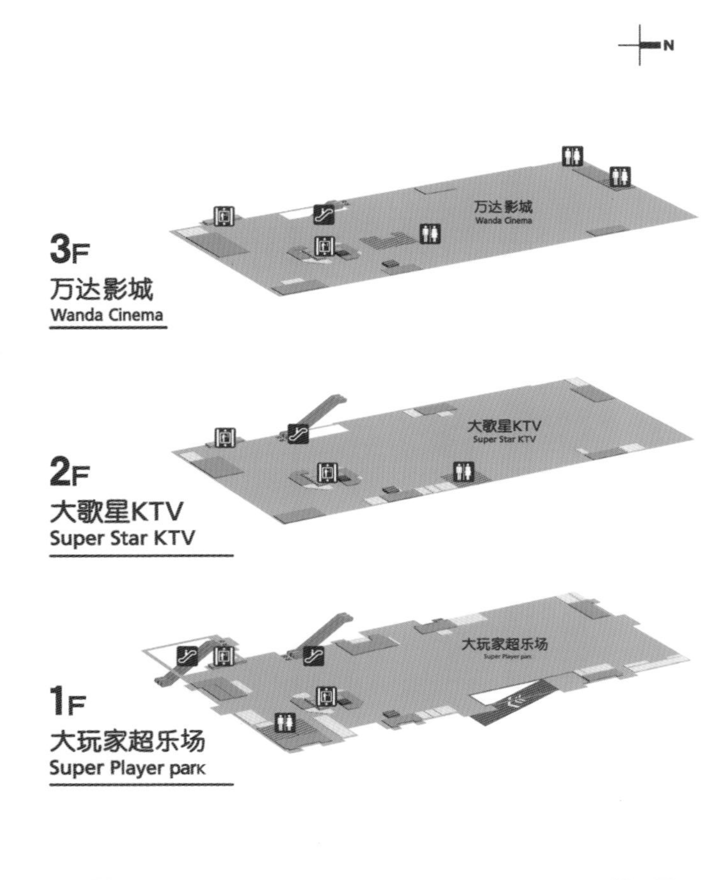

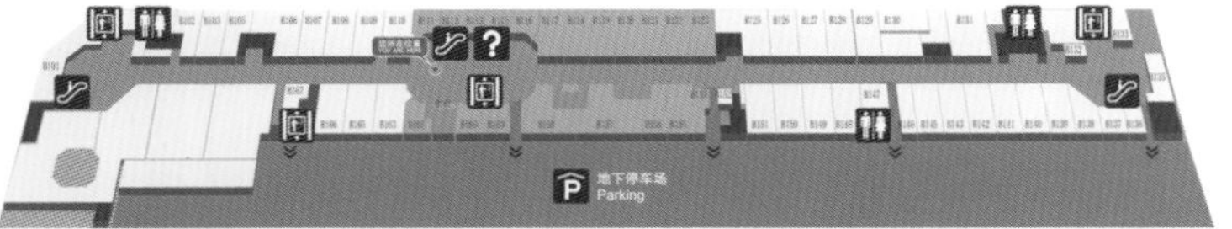

B1
室内步行街
Pedestrian Street

品牌落位图

室内步行街中庭

室内步行街

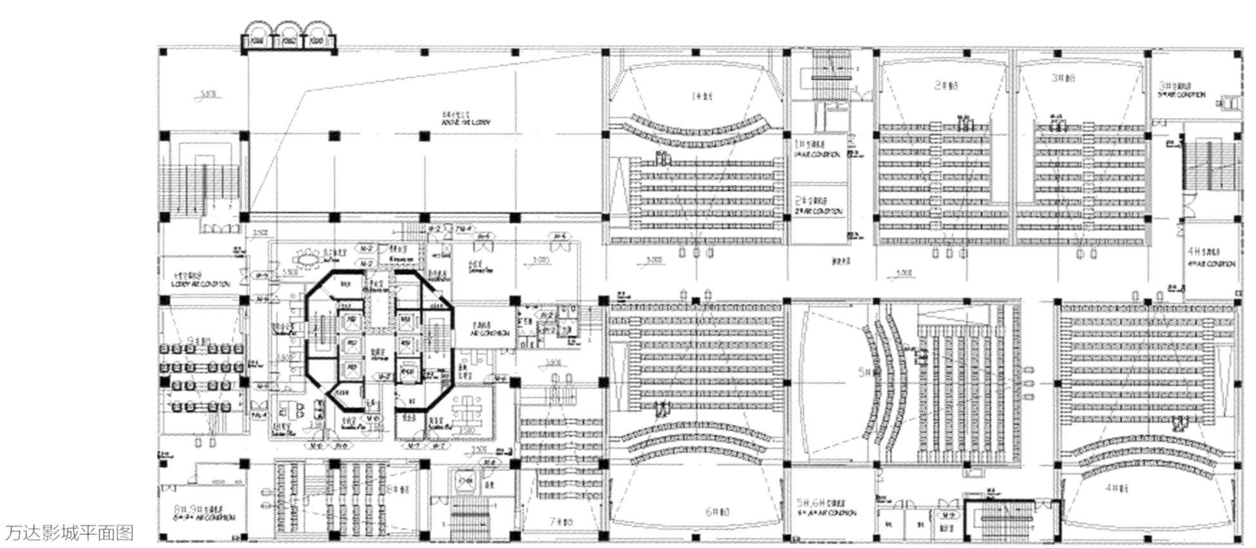

万达影城平面图

室内步行街

室内步行街中庭

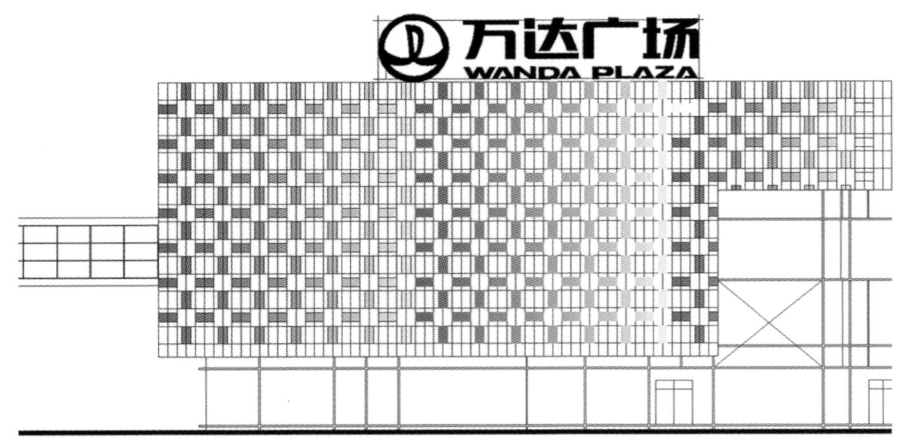

广场局部立面

广场夜景

重庆南坪万达广场
CHONGQING NANPING WANDA PLAZA

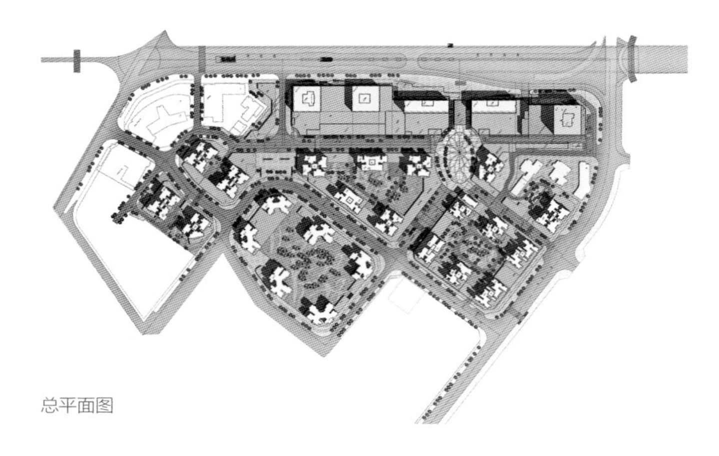

总平面图

重庆南坪万达广场位于重庆市南岸区，在重庆市东南部，项目所在区域为重庆市五大商圈之一的南坪商圈的中心。重庆南坪万达广场总用地面积 9.84 公顷，总建筑面积 78 万平方米，由商业中心、五星级酒店、写字楼、高级公寓及高端住宅组成，是万达城市综合体的代表之作，无论在设计上，还是在主力店的组合上都代表了重庆商业中心的最高水准。同期开业的超五星级万达艾美酒店是重庆最好的酒店之一，万达艾美酒店的开业打破了南重庆无超五星级酒店的历史，大幅提升重庆南岸的区域形象和商务品质。

Chongqing Nanping Wanda Plaza is located in NanAn District, southeast of Chongqing. The project site is in the center of Nanping commercial circles, one of the five great in Chongqing. The site area is 98,400 m^2, whole building area is 780,000 m^2. Consisting of shopping center, five-star hotel, office tower, upper-class apartments and Wanda mansions as well, this urban complex is the representative of Wanda Plaza and the highest level both in architecture design and the type selection of anchors. In the same time Le Méridien opens to be one of the best hotels in Chongqing. The operation of Le Méridien changes the history of absence of super five-star hotel and greatly improves Nanan District's image and quality.

室外步行街

重庆南坪万达广场外立面

重庆南坪项目主入口

室内步行街大中庭

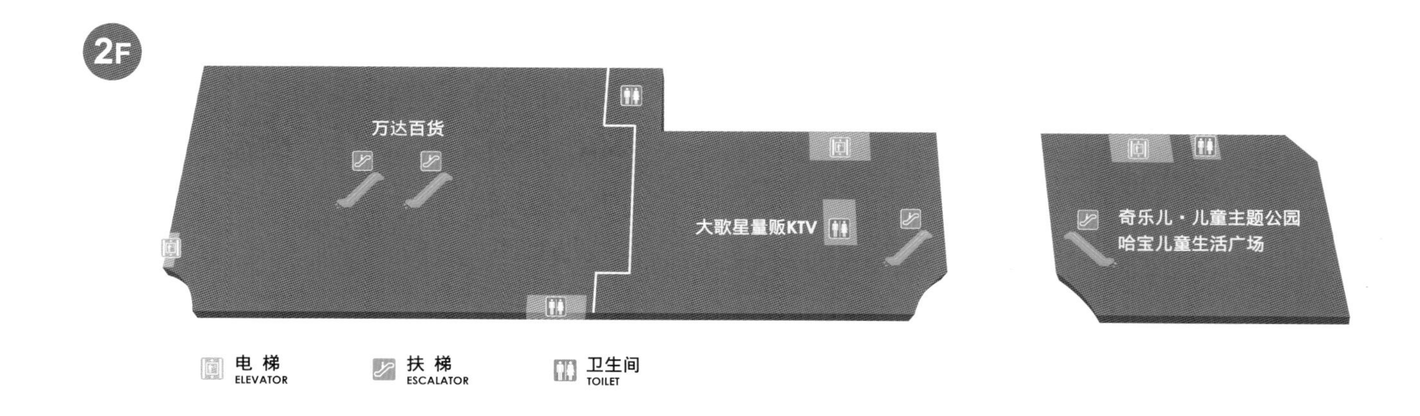

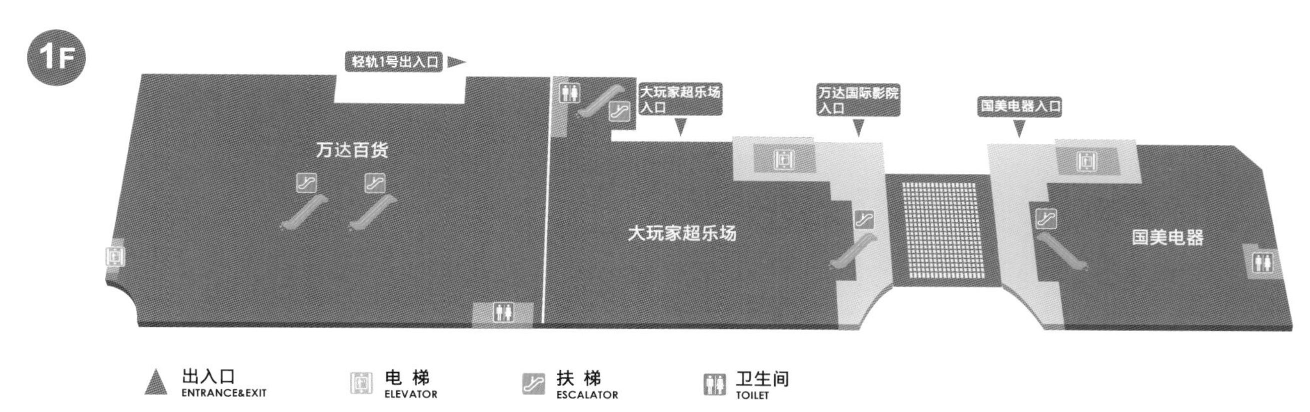

品牌落位图

万达影城平面图

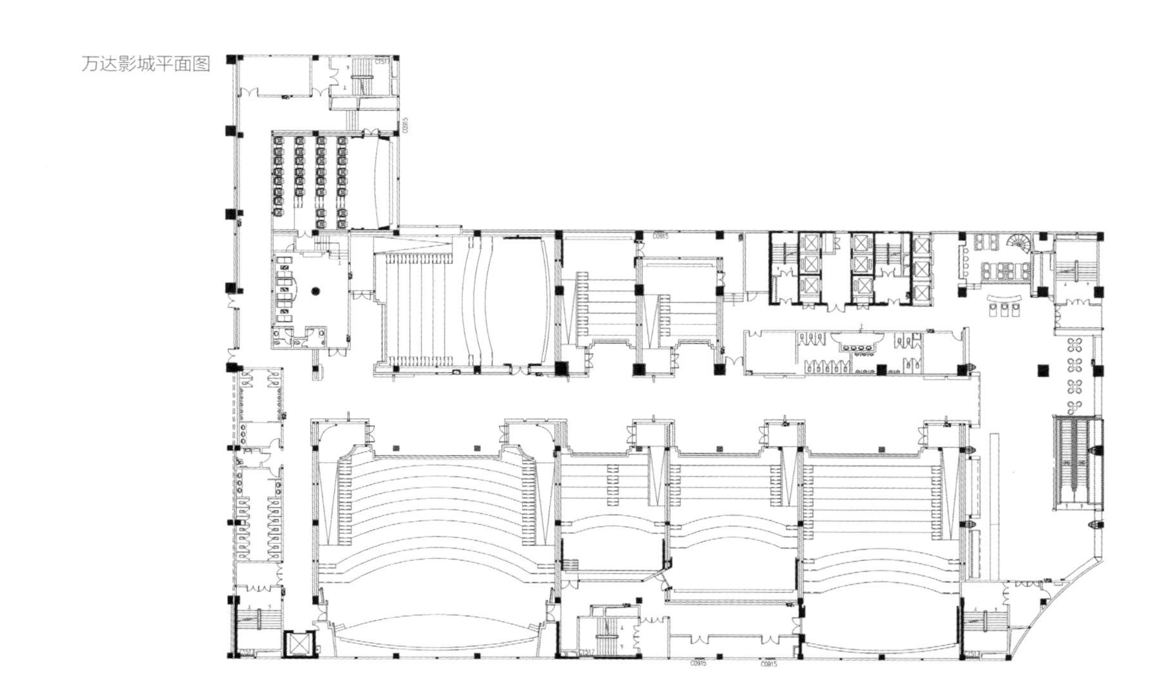

Createa
佰草集
HERBORI
CASIO
宜北町
Createa

室内步行街大中庭

西安民乐园
万达广场
XI'AN MINLE PARK WANDA PLAZA

室内步行街中庭

西安民乐园万达广场位于西安市新城区民乐园地区，东五路以南，东新街以北，解放北路以东和尚勤路以西。总规划用地面积为11.99万平方米，其中建设用地7.79万平方米，其他代征用地4.2万平方米。规划总建筑面积约28.99万平方米，其中地上部分22.9万平方米，地下部分6.09万平方米。

Xi'an Minle Park Wanda Plaza is located in the Minle Park, Xincheng District, Xi'an, with Dongwu Road to its south, Dongxin Street to its north, North Jiefang Road to its east and Shangqin Road to its west. The plaza covers a total area of 119,000 m^2, of which 77,900 m^2 is construction area and 42,000 m^2 public land. The planned gross floor area is 289,900 m^2 which consists of 229,000 m^2 above-ground area and 60,900 m^2 underground area.

商业综合体是西安民乐园万达广场的核心物业，以"天街"、"天庭"、"天梯"为主要特色，主力店业态包括：万达百货、超市、万达影城、大歌星KTV、大玩家超乐场、电器城、儿童用品商店、酒楼等。

The commercial complex is the core property in Xi'an Minle Park Wanda Plaza, which's features is "Heaven Street", a "Heaven Atrium" and a "Heaven Ladder". The main programs of the anchor stores include Wanda Dept. Store, supermarket, Wanda Cinemas, Super Star KTV, Super Player Park, electronics mall, children's goods store and restaurant.

室内步行街中庭

设计特色：
Design features:

“天街”：创造性地将室内商业步行街布置在购物中心的三、四、五层，面积达 4.3 万平方米，营造西安最具时尚的购物环境。

“天庭”：在商业综合体的主入口处规划设计 25 米高、面积达 1500 平方米、5 层贯通的巨型中庭，创造西安最大的商业共享空间。

“天梯”：设置西安当地首台提升高度超过 20 米，由首层直达五层的双向扶梯。

"Heaven Street": creatively placing the 43,000 m^2 indoor shopping street on the third, fourth and fifth floors of the shopping mall to create the most fashionable shopping environment in Xi'an.

"Heaven Atrium": boldly placing a 25-meter tall, 1,500 m^2, 5-story huge atrium at the entrance of the commercial complex to create the largest commercial public space in Xi'an.

"Heaven Ladder": installing the first two-way 20-meter-plus elevator which links from ground floor directly to the fifth floor in Xi'an.

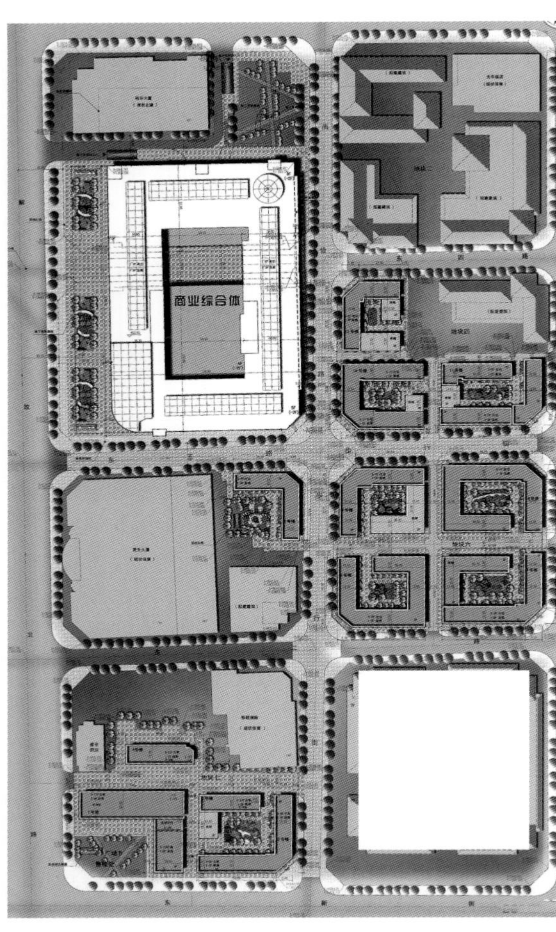

总平面图

广场西南角夜景

西安站
15
GXG
GILL MIX GREEN
4F
3F
歌莉

室内步行街

广场东北角日景

广场西南角夜景

连廊下方景观夜景

天庭外夜景

广场入口夜景

步行街中庭

中庭

1F

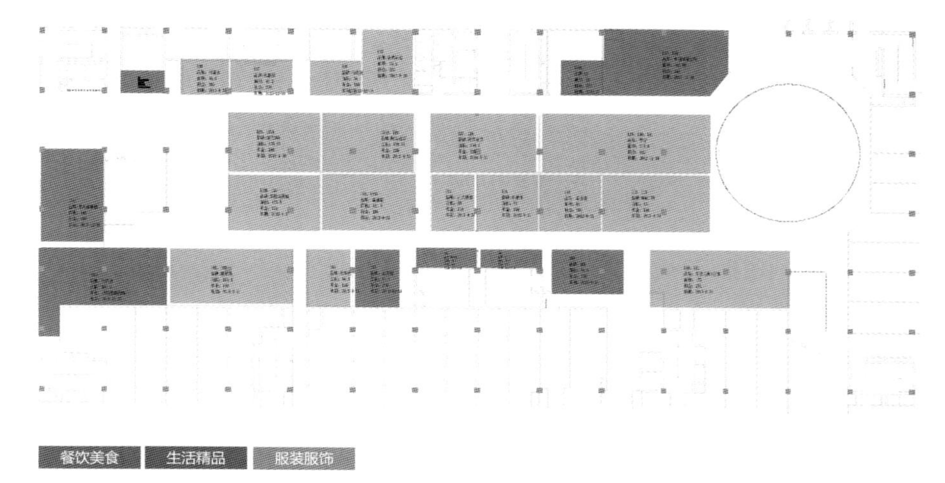

品牌落位图

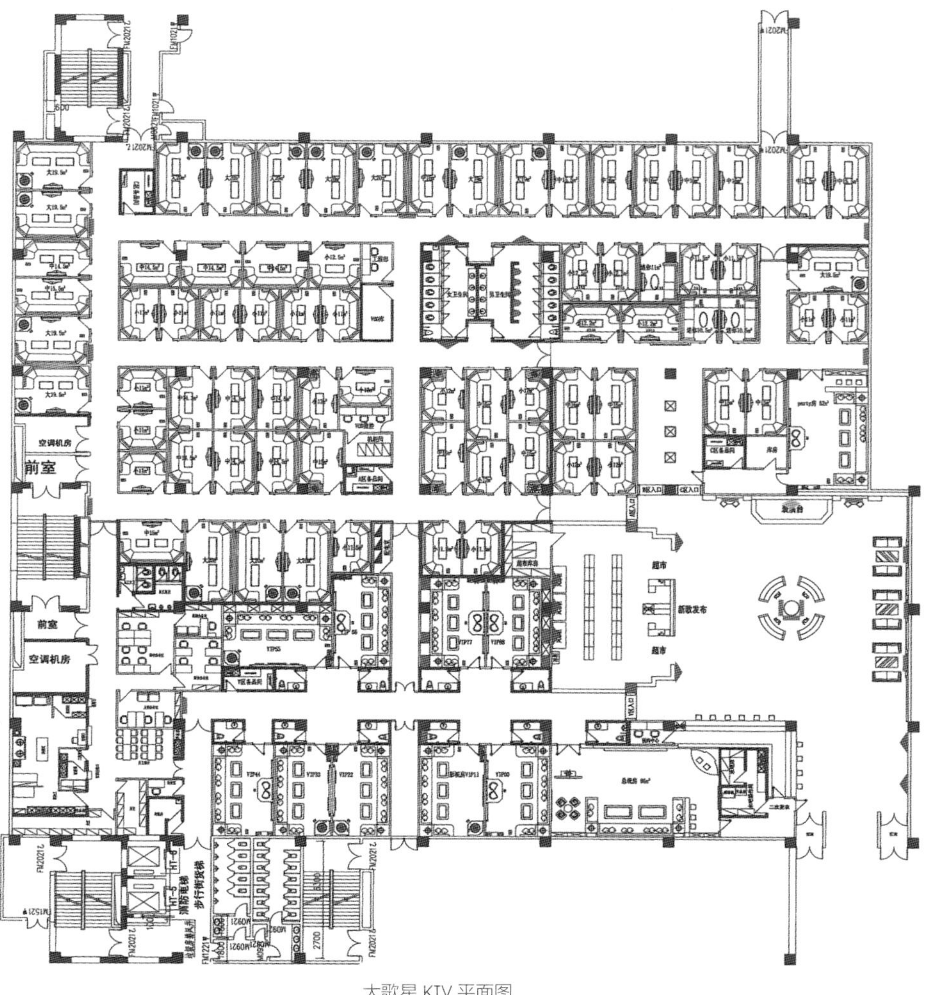

大歌星 KTV 平面图

洛阳
万达广场
LUOYANG WANDA PLAZA

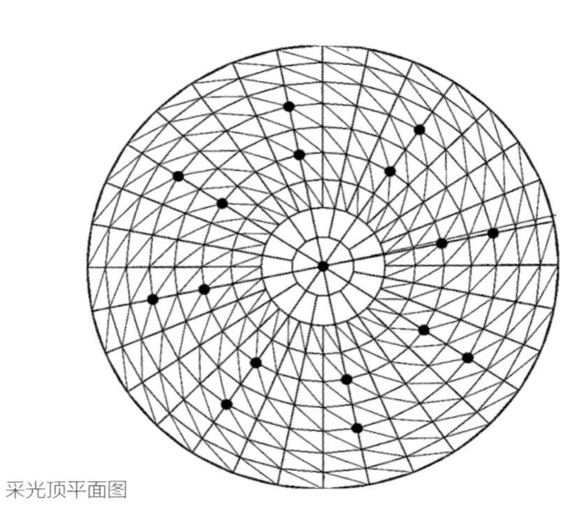

采光顶平面图

洛阳万达广场位于洛阳市涧西区。地块范围东至珠江路，南至辽宁路，西至丽新路，北至洛阳少年文化中心南界。规划用地面积 3.24 万平方米，其中代征用地 0.6 万平方米。

Luoyang Wanda Plaza is located in Jianxi District, Luoyang City. The plot ranged from east to the Zhujiang Road. south to the Liaoning Road. west to the Lixin Road, north to the south side of Luoyang Youth Cultural Center. The planning land area is 32,400 m^2, in which generation of commandeer land: 6000 m^2.

用地是由近等边三角形用地以及长条形用地构成。因用地的限制及限高的要求，在三角形用地规划了商业广场，沿长边设计了室内步行街；在条形用地上规划了室外步行街。

Land was composed of square near equilateral triangle and a strip shape. Because of the limit of land use and high requirements, in the triangle of land use planning commercial plaza, along the long edges of the interior design of pedestrian streets; on the strip land an outdoor pedestrian streets has been planned.

室内步行街中庭

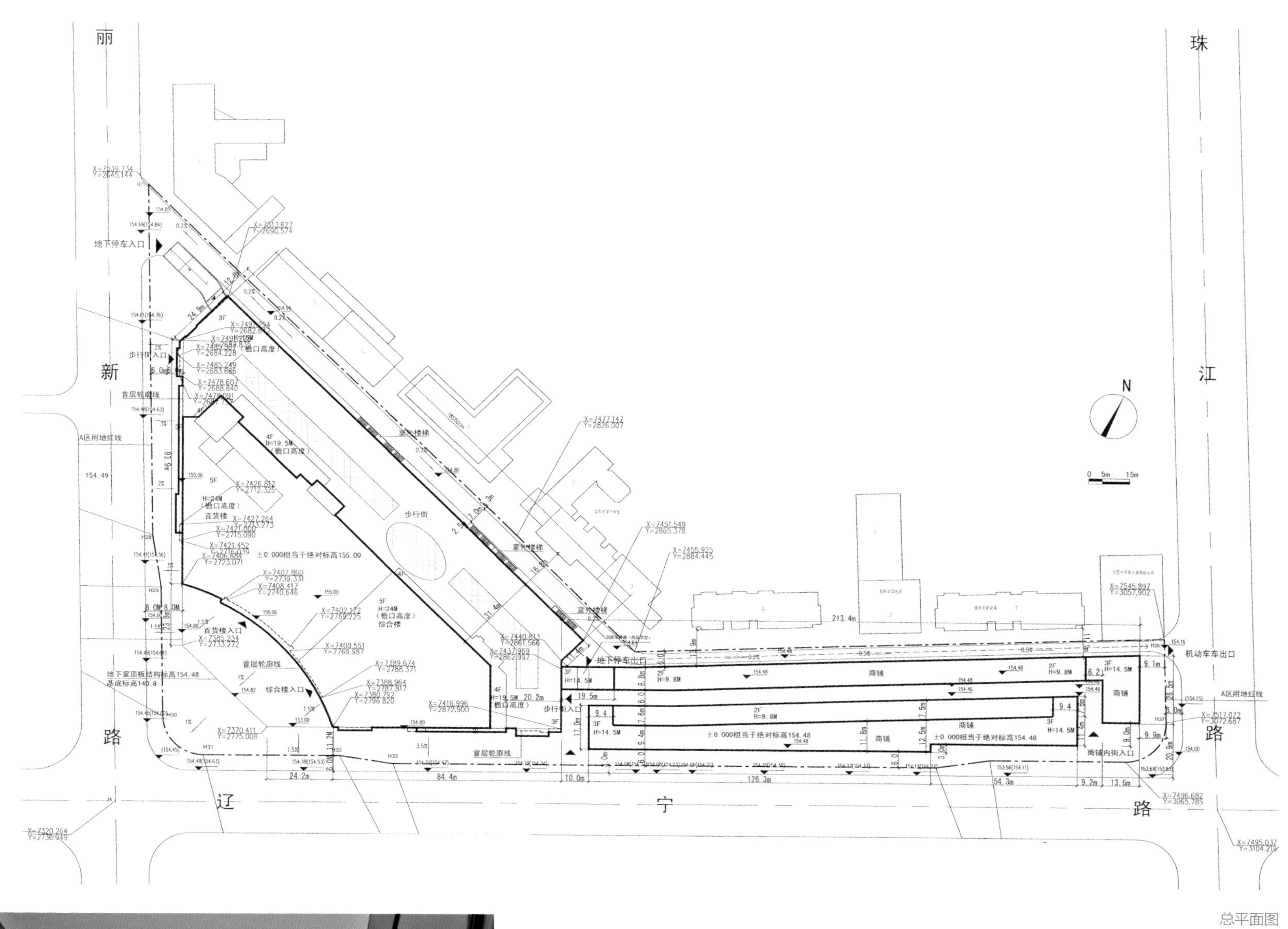

总平面图

室外步行街夜景

本项目由商业广场及室外步行街组成，总建筑面积11万平方米（其中地上建筑面积6.5万平方米，地下建筑面积4.5万平方米）。建筑高度24米；层数8层，其中地上5层，地下3层。规划设计车位512个。

The project is composed of the commercial square and outdoor pedestrian street. The total construction area is 110,000 m^2 (including ground area of 65,000 m^2, area of 45,000 m^2 underground). The building's heighth is 24m; with 8 layers, wherein the ground 5 layers, underground 3 layers. The planning number of parking spaces is 512.

广场日景

1F:时尚名品

服装服饰
生活精品
餐饮美食

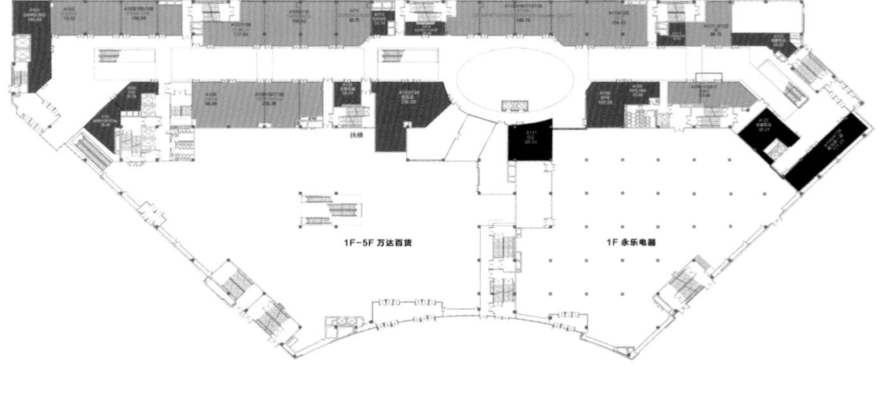

品牌落位图

2F:风尚潮流

- 服装服饰
- 生活精品
- 餐饮美食

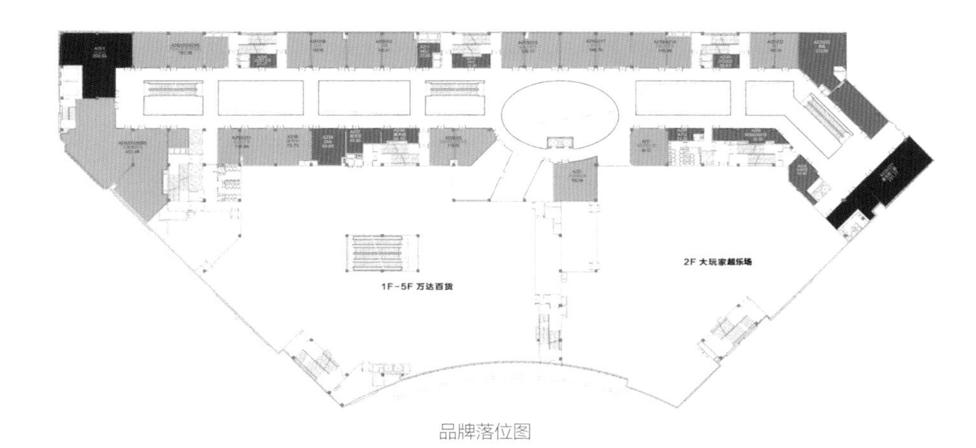

品牌落位图

广场夜景

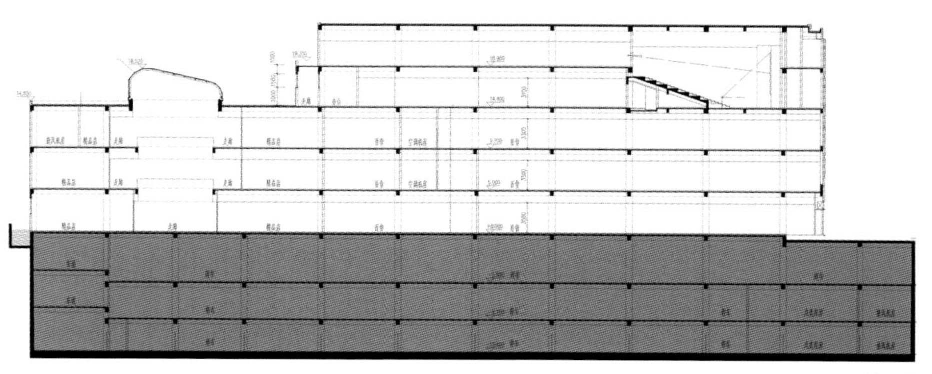

剖面图

室内步行街

万达影城入口

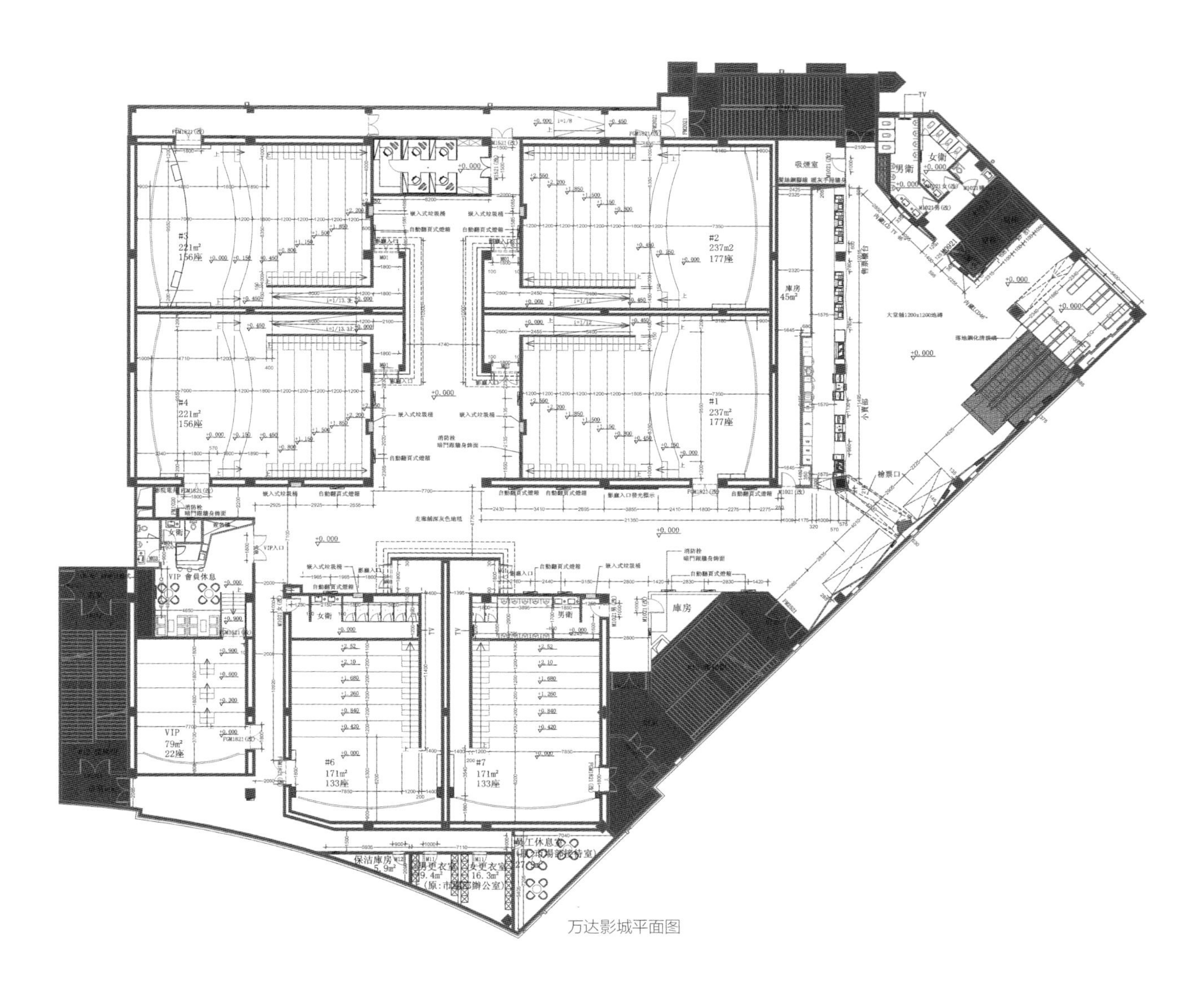

万达影城平面图

PART 2 酒店 HOTELS

万达集团目前已开业 28 家五星和超五星级酒店。万达集团计划到 2015 年开业 80 家五星和超五星级酒店，营业面积达 300 万平方米，成为全球最大的五星级酒店业主。万达拥有五星级酒店品牌——万达嘉华、超五星级酒店品牌——万达文华以及顶级奢华酒店品牌——万达瑞华。

Wanda Group has opened 28 five-star and super five-star hotels so far. The Group also plans to open 80 five-star and super five-star hotels by 2015, with an operation area of 3 million m^2, as the world's largest five-star hotel owner. Wanda Group owns its five-star hotel brand—Wanda Realm, super five-star hotel brand—Wanda Vista and top luxury hotel brand—Wanda Reign.

万达集团旗下酒店分为城市类酒店和度假类酒店两大类。万达城市类酒店通常位于城市中心，一般和万达广场等业态组成万达城市综合体。万达度假类酒店通常位于文化旅游区，一般和旅游小镇、秀场、主题公园等组成万达文化旅游项目。

Wanda Group's hotels are divided into two major categories: the urban hotels and the resort hotels. The urban hotels are usually located in the city center, normally forming a Wanda urban complex together with Wanda Plaza, etc..The resort hotels are usually located in the cultural tourism zone, normally forming a cultural tourism project together with the tourist town, the show theatre, the theme park, and so on.

青岛万达
艾美酒店
LE MERIDIEN QINGDAO

青岛万达艾美酒店外景

青岛万达艾美酒店距离青岛流亭国际机场仅 30 分钟车程，距青岛著名景点——啤酒街、奥帆中心及海滩等仅 10 分钟车程。总建筑面积 4.6 万平方米，拥有 321 套客房。青岛万达艾美酒店以宽敞时尚的客房及会议空间、世界级美食以及独具特色的服务，为来自世界各地的客人们开启了富含创意及趣味的青岛之旅。青岛万达艾美酒店的设计极致时尚优雅，奢华浪漫的装饰风格、风情韵味的灯效光影、华美璀璨而又无处不在的巨大水晶灯，无不一一展示着万达艾美的与众不同。

Le Meridien Qingdao is located in a distance of only 30 minutes by car from Qingdao Liuting International Airport, and only 10 minutes by car from well-known scenery spots of Qingdao, such as Beer Street and Olympic Sailing Centre and the beach. It has an overall floorage of 46,000 m^2, with 321 guest rooms. With a spacious and fashionable guest rooms and conference space, world-class food and unique service, the hotel starts an creative and interesting travel for tourists from all the world. The design of the hotel is delicate, fashionable and elegant, with a luxurious decoration style and amorous feeling effects of light and shadow, huge beautiful and bright crystal lights everywhere. Everything presents the unique of Wanda Le Meridien.

酒店入口夜景

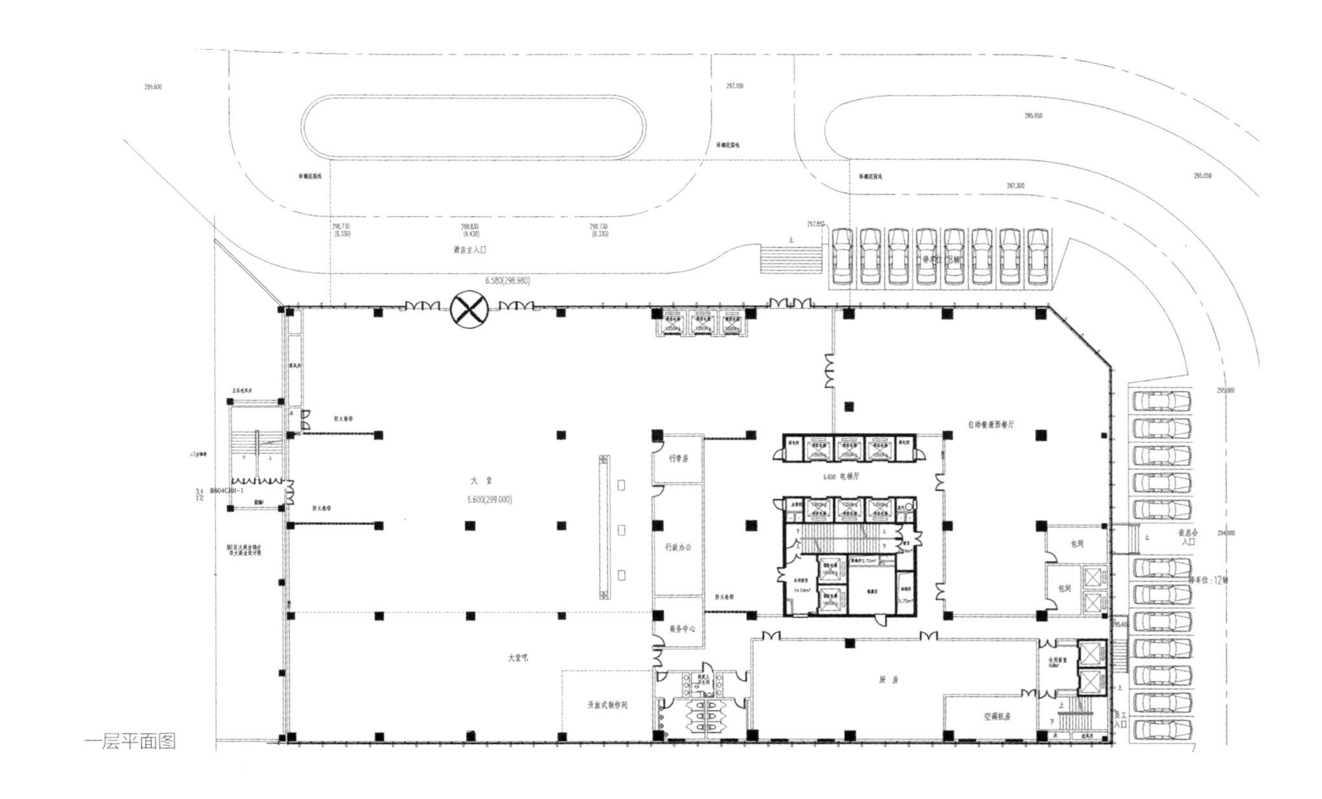

一层平面图

Le MERIDIEN
万达广场
WANDA PLAZA
Le MERIDIEN 青岛万达艾美酒店
海泊河
HAI BO HE
美食舰队
MEISHIJIANDUI

酒店夜景

MERIDIEN
万达
U.B.C.

酒店大堂

酒店套房

电梯间

室内游泳池

酒店大堂局部

酒店大堂局部

重庆万达艾美酒店

LE MERIDIEN CHONGQING, NAN'AN

酒店门头夜景

酒店夜景

Le Meridien Chongqing Nan'an locates in No.10 Jiangnan Rd. Nan'an District in Chongqing, neighboring newlybuilt Wanda Plaza with housing, shopping and entertainment center. The hotel lies in the river front with very convenient transportation, and is about 15 min's walking distance to Changjiang, and 10 min's driving distance to CBD of Jiefangbei. The design of Le Meridien reflects this city's modernism and fashion, and in the meantime integrates the local culture and tradition features as well. The hotel site area is 5,000 m^2, gross building area is 43,000 m^2. The main building of hotel is a 23-storey tower with 319 guest rooms. The interior is spacious and bright, its style is simple but elegant, and its landscape is designed relaxing and pleasant to rescue guests from busy business routines. The fitness center and indoor swimming pool are available on site too. Considering the request of various conference and ceremony, the hotel is prepared with many conference rooms and multipurpose halls.

重庆万达艾美酒店坐落于重庆市南岸区江南大道 10 号，与重庆最新兴的住宅、购物与娱乐区——重庆南坪万达广场毗邻，地处长江之畔，交通便利，步行 15 分钟即抵长江，距解放碑中央商务区仅 10 分钟车程。酒店设计风格体现了城市的现代与时尚，同时也融合了当地的文化与特色。酒店占地 0.50 公顷，总建筑面积 4.3 万平方米；主楼高 23 层，有各式客房与套房 319 间（套），空间宽敞明亮，风格简约雅致，环境轻松宜人，将宾客从繁忙的观光游览和商务日程中解脱出来。酒店的健身俱乐部和室内泳池等休闲设施齐全，同时酒店设有会议室和多功能厅，是宾客举办各种会议、庆典活动的理想场所。

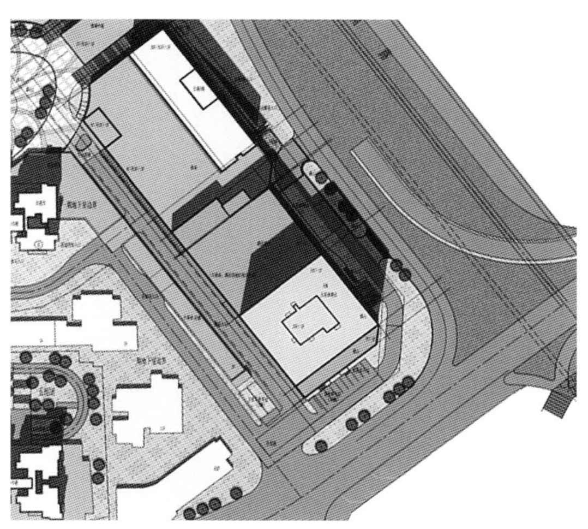

总平面图

酒店主入口日景

酒店大堂

客房

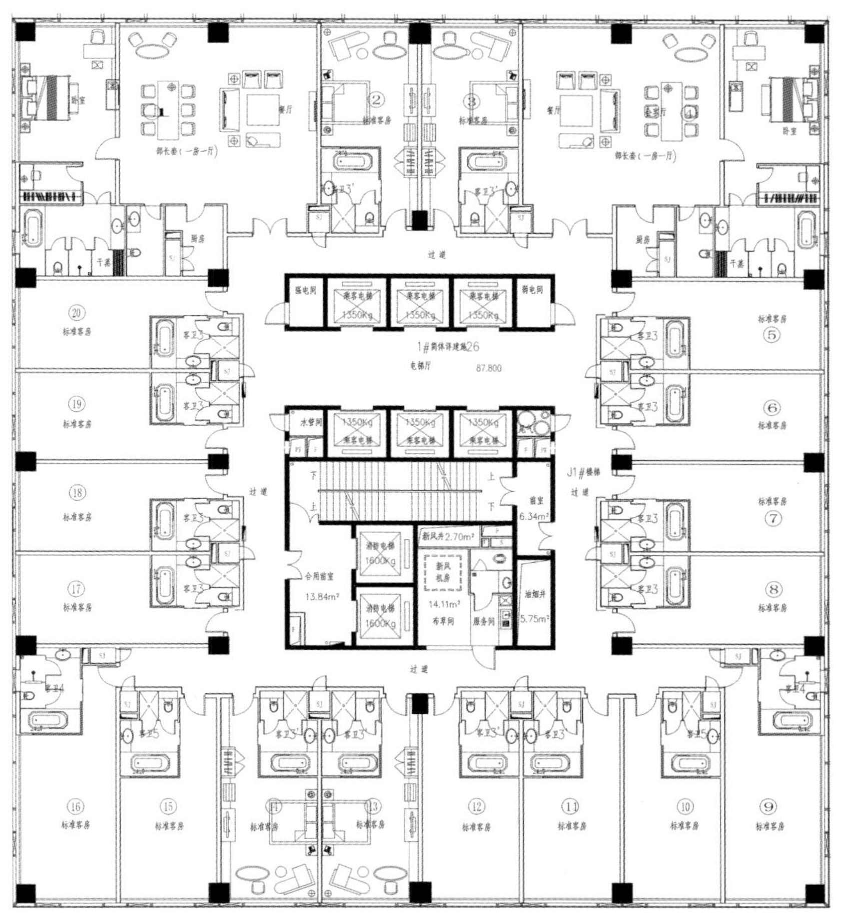

标准层平面图

酒店室内泳池

PART 3 导向标识精选 SELECTION OF GUIDING SIGNS

导向标识系统是引导人们在公共场所活动的综合性公共信息系统。它的基本功能是指引方向，重要辅助功能是强化区域形象。

Way-finding & signage system is the comprehensive public information system that is used to guide people's activities at public places. Its basic function is to guide the direction, with an important auxiliary function of enhancing the image of the region.

万达广场的导向标识系统经过多年的项目实践及总结，已经形成自身较为完善的一套规划设计标准，其中主要包括户外、室内和地下停车场三大部分。

After numerous years of project practice and summarization, the way-finding & signage system of Wanda Plaza has become a complete set of planning and designing norms, mainly including three parts: the outdoor, the indoor and the underground parking space.

设计中标识布点须遵循车、客流行进流线，并结合现场环境设置，保证标识的醒目、易识别和场地的视觉通透性。

In designing, the signage layout must follow the routes of vehicles flow and passengers flow, and, based on the environmental conditions, guarantee that the signage is eye-catching and easy to recognize, providing visual permeability to visitors.

不同牌体的造型、设计元素、牌体本身的尺度及材质需与广场对应区域的整体设计风格相协调一致并满足各功能性信息的排布要求。

The shape, designing elements, dimension and material quality of different signs shall conform to the overall designing style of the corresponding areas of the Plaza and satisfy the requirement for the configuration of all kinds of functional information.

万达广场
WANDA PLAZA
SELECTED

公寓
APARTMENT
写字楼A·B座
TOWER A, B
万达铂尔曼大饭店
PULLMAN
室内步行街
PEDESTRIAN STREET
万达广场
写字楼C、D座
TOWER C, D
万千百货
VAN'S
运动100
SPORT 100
国美电器
GOME
家乐福
CARREFOUR
大玩家超乐场
S-PLAYER
大歌星KTV
SUPER STAR KTV
万达国际影城
WANDA CINEMAS

2F
3F 万千百货
大歌星量贩KTV
万达影城
2F 万千百货
国美电器
大玩家超乐场
1F 万千百货
UNIQLO
ZARA
B1 华润苏果
地下停车场
B2 地下停车场

2F

3F
万达国际影城
酒店
良友金都美食城
LIANGYOU FOOD COURT
万千百货
2F
大歌星KTV
SUPER KTV
酒店
奇乐儿亲子园
CHEER KIDS PLAY PARK
万千百货
1F
大玩家超乐场
酒店
国美电器
GOME
万千百货
佳世客超市
JUSCO
乐世界

客梯
1F
德庄火锅 4F
菜香源酒楼
圆缘园时尚餐厅 3F
全汉斯南美烤肉
奇乐儿·儿童主题公园 2F
哈宝儿童生活广场
国美电器 1F
国美电器 UG
室内步行街 LG
(缤纷大地)
室内步行街 B1
食尚大道
地下停车场
客梯

2F
4F
海底捞火锅
HAIDILAOHUOGUO
汉拿山韩式烧烤
HAN NA SHAN KOREAN RESTAURANT BBQ
3F
食赏坊
SNACK STREET
红黄蓝亲子园
R.Y.B LEARNING CENTER
奇乐儿儿童主题公园
CHEER KIDS PLAY PARK
2F
魅力街
STYLE STREET
国美电器
GOME
1F
名品荟
FASHION STREET
运动100
SPORT 100
国美电器
GOME
B2
地下停车场
PARKING

万千百货
Van's Dept.Store
盛大开业
GRANDOPEN
时尚启城

万千百货
万达影城
大中电器
大歌星KTV
家乐福超市
时尚旅酒店

万千百货
VAN'S DEPT STORE
国美电器
GOME
2
大玩家超乐场
SUPER PLAYER PARK

Information
UNIQLO
ZARA

金沙江西路
1号门
万千百货

无印良品
MUJI
1F
活力万达
时尚启城
盛大开业
GRAND OPEN
万达广场
WANDA PLAZA

3F
食赏坊
SNACK STREET
2F
魅力街
STYLE STREET
1F
名品荟
FASHION STREET
万达广场

P
地下停车场入口
万达广场

万达广场
WANDA PLAZA

万达广场
WANDA PLAZA
万千百货
VAN'S DEPT.STORE
喜来登酒店
Sheraton Hotel
时尚旅酒店
SMART Hotel
办公楼B-C
Office Block B-C
2号门
Gate 2
金街南口
Finance Street (S)

康莱德酒店
CONRAD HOTEL
希尔顿酒店
HILTON WANDA
万达中心
写字楼
WANDA CENTER
OFFICE
万达中心
WANDA CENTER

INFORMATION
总信息
泰州万达广场平面图
THE PLAN OF TAIZHOU WANDA PLAZA
金街
Golden Street
酒吧街
Bar Street
时尚旅酒店
Smart Hotel
万达希尔顿
逸林酒店
Doubletree By Hilton

Information
购物中心2号门
SHOPPING CENTER GATE 2
万千百货
VAN'S DEPT.STORE
UNIQLO
ZARA
时尚旅酒店
SMART HOTEL
写字楼A/B座
Tower A/B
公寓1号楼
Apartment Building 1
汽车客运站
城市候机楼
Information
镇江万达广场平面图
THE PLAN OF ZHENJIANG WANDA PLAZA

PART 4 2009年以前开业项目

PROJECTS BEFORE 2009

南昌八一
万达广场
NANCHANG BAYI WANDA PLAZA

2002 年开业

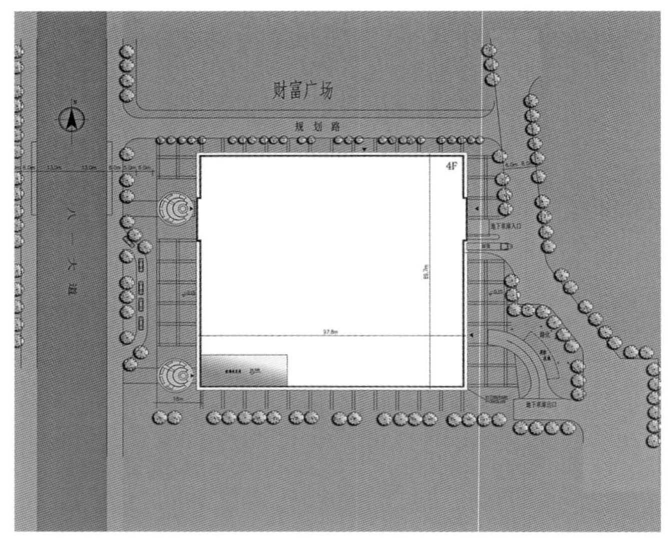

总平面图

南昌八一万达广场地处南昌市 CBD 核心地带，南临八一广场，西临八一大道、北临财富广场、东临广场北路，交通便利，人气旺盛。总用地面积 1.4 万平方米，总建筑面积达 5 万平方米，广场共 5 层，其中商业主楼 1~4 层，地下一层为大型停车场，首层为大气典雅的名品专卖店，二、三层为沃尔玛购物广场，四层为万达影城。整个广场集时尚购物、休闲美食、文化娱乐为一体，业态合理，布局规整，为南昌市民和广大消费者提供了一站式购物的主题购物中心。

Nanchang Bayi Wanda Plaza is situated in the core of CBD of Nanchang. It is closed to Bayi Square to the south, Bayi Road to the west, Caifu Square to the north, and Guangchang North Road to the east, with convinient transportation, which make it very popular. It has a total floor area with 14,000 m^2 and an overall floorage of 50,000 m^2. There are a total of five storeys in the plaza. Commercial main building is located in the first to fourth floor. It has a large car park in the ground floor, a generous and elegant brand store in the first floor, Wal-Mart shopping mall in the third floor, and Wanda Cinemas is located in the fourth floor. The whole plaza collects fashion shopping, leisure and food, culture and entertainment as a whole. With reasonable business forms and a regular layout, the plaza provides an one-stop shopping themed centre for Nancang residents and all the customers.

省妇幼保健院
大玩之家
万达国际电影城
万达购物广场
WAL★MART
SUPERCENTER
沃尔玛购物广场
水饺

广场东南角鸟瞰

上海五角场
万达广场

SHANGHAI WUJIAOCHANG WANDA PLAZA

2006 年开业

广场日景鸟瞰

项目位于上海市杨浦区，东临淞沪路、南靠邯郸路、西邻国宾路、北依政通路，规划用地 6.0 万平方米，总建筑面积 33.40 万平方米，总投资 30 亿元人民币，是万达进入上海的第一个商业项目，是万达的第三代商业综合体。

项目规划由地下步行街、商业组合体、商务办公等业态构成。2006 年 12 月开业，汇集了沃尔玛超市、巴黎春天百货、大玩家超乐场、万达影城、黄金珠宝、大歌星 KTV、宝大祥、第一食品、特力和家居馆、上海书城等十大主力店，以及近百家各具特色的精品餐饮休闲品牌，成为该地段的城市商业中心。

The project is located in Yangpu District. The Plaza has it's major frontage, east to the songhu road, south to Handan road west to the Guobin road and north to the Zhengtong road. The Planning area is 60,000 m^2, with a total construction area of 334,300 m^2, a total investment of RMB 3 billion. It is the first commercial project when Wanda enters into Shanghai, and it also Wanda's 3rd Generation Commercial Complex.

Project planning is composed of the underground pedestrian street, business combination, business office and other formats. It opened in Dec. 2006. It has a collection of Wal-Mart supermarket, Paris spring department store, Super Player Park, Wanda Cinemas, gold jewelry, Super Star KTV, Daxiang, the first food, hola home furnishing, Shanghai Bookstore, the ten main store. Nearly 100 unique products specially catering leisure brand, which becomes the business center of this area.

总平面图

SAMUEL & KEVIN
玛花纤体
特力时尚汇3F
3362-1322
NEW BODY NEW LIFE
LIFE 1 PLAZA
HOLA 特力屋
韩林炭烤
1F/2F流行服饰、时尚专区 3F精品生活馆
WALL STREET ENGLISH
华尔街英语
特力时尚汇
广场夜景

咨询 65647786
新东方
BOSHIWA
玛花纤体
广场日景

巴黎春天
650万 圆梦中环墅
5067 6818
巴黎春天

广场日景

万达影城 IMAX
CITYLIFE
365博士蛙
西贝莜面村
TASTY
望湘园
角场黄金珠宝

FANCL
AGATHA
AGATHA
AGATHA
AGATHA C'EST MOI
AGATHA C'EST MOI
AGATHA C'EST MOI

室内步行街中庭

3F

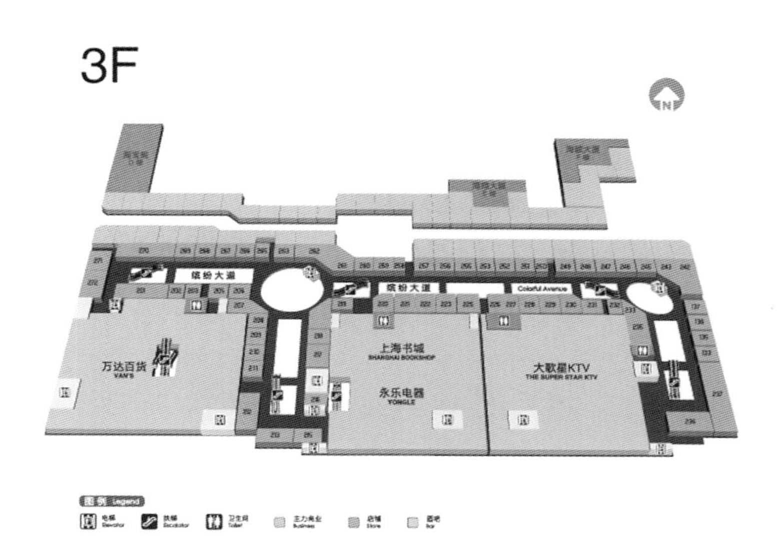

2F

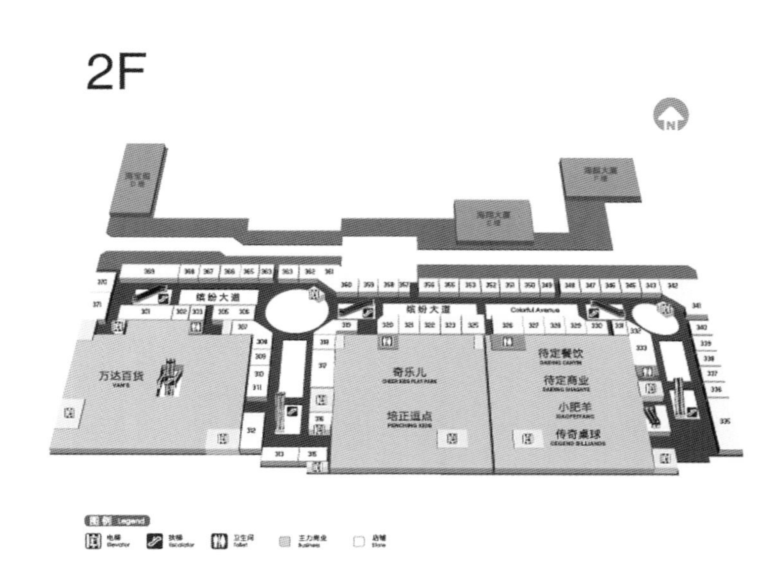

1F

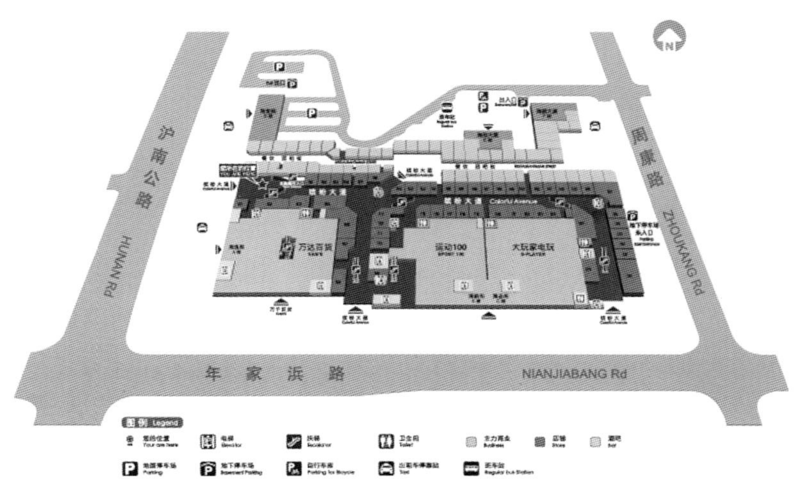

品牌落位图

北京CBD万达广场
BEIJING CBD WANDA PLAZA

2006 年开业

影院大厅

北京 CBD 万达广场位于北京市中央商务区核心地段，紧邻东三环和国贸中心，坐北朝南，占地面积约 10 公顷，总建筑面积约 45 万平方米；南区由三栋面朝东长安街颇具雕塑感的塔楼组成，北区共有 12 栋高层板式写字楼，南北区共同构成 CBD 万达广场。

Beijing CBD Wanda Plaza is located in the core CBD of Beijing, next to East 3rd Ring Road and China World Trade Center. The south-facing plaza covers a total area of 100,000 m^2, with a gross floor area of 450,000 m^2. The plaza consists of a north section and a south section: The south section includes three sculptural towers facing East Chang'an Avenue. The north section includes twelve plate-style office towers.

严谨而理性的建筑线条，简洁而隽永的设计语言，使万达广场与西侧的国贸中心、东侧的华贸中心连接起来，相映生辉，以不同的建筑形象共同刻画出东长安街美丽的天际线。

The carefully-designed patterns, simple but powerful design language of the Wanda Plaza corresponds to the China World Trade Center and Huamao Center and together they form a beautiful skyline for East Chang'an Avenue with different but harmonious architectural images.

广场日景

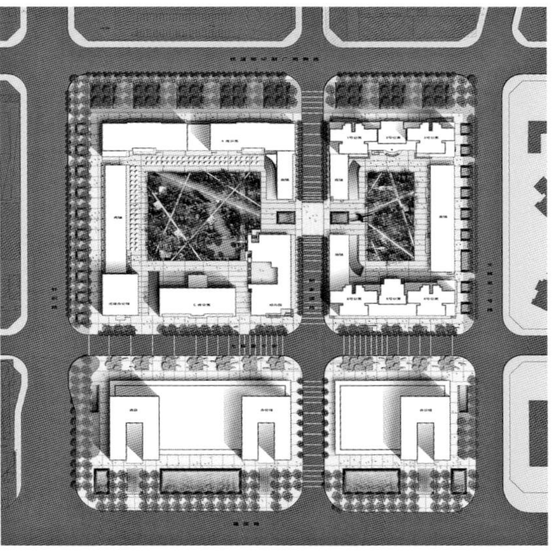
总平面图

广场鸟瞰

广场鸟瞰

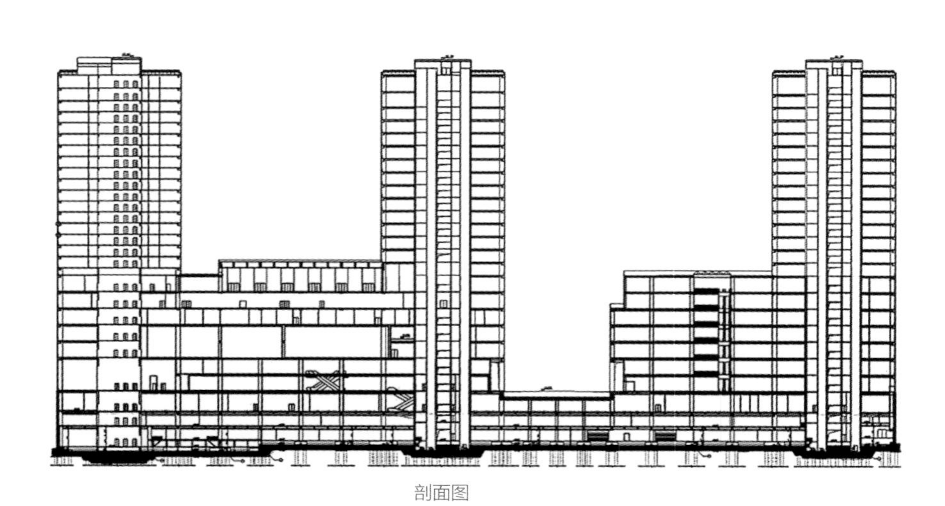
剖面图

万达国际电影城
WANDA INTERNATIONAL CINEMAS
NEW WORLD DEPARTMENT STORE
SUPERCENTER
沃尔玛
WANDA PLAZA

SOFITEL
WANDA
万达广场
万达广场
万达国际电影城
WANDA INTERNATIONAL CINEMAS

广场夜景

广场鸟瞰

宁波鄞州
万达广场

NINGBO YINZHOU WANDA PLAZA

2006 年开业

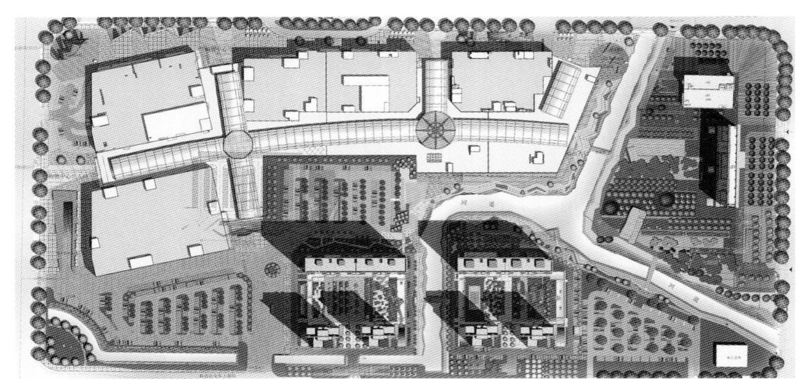

总平面图

广场鸟瞰

外立面夜景照明

宁波鄞州万达广场位于宁波鄞州区中心，东面紧临城市主干道天童北路，北靠四明中路，南依贸城中路，毗邻鄞州区政府，西至宁南北路，总占地面积 21.09 公顷，是由大型商业购物中心、五星级酒店、超高层城市公寓三大功能板块构成的大型城市建筑综合体。宁波鄞州万达广场属万达第三代商业地产的开山之作，首次引入了室内步行街，在万达商业地产的发展中具里程碑作用。国际购物中心位于万达广场西北部，总建筑面积近 26 万平方米，是迄今为止万达开发的最大规模的商业广场，包含了大型国际连锁超市、建材超市、数码广场、娱乐广场、电影美食广场、家居广场、国际百货等多种商业形态，汇集了沃尔玛、百安居、HOLA 特力屋、喜百度、苏宁电器、石浦大酒店、大玩家超乐场、大歌星 KTV、万达影城、银泰百货、吉盛伟邦家居等 11 家主力店及其他近 200 家境内外知名品牌店。宁波鄞州万达广场商圈已发展成为宁波第二大商圈。

Ningbo Yinzhou Wanda Plaza is located in the centre of Yinzhou District of Ningbo. It is closed to high street of Tiantong North Road in the east, Siming Middle Road to the north, Maocheng Middle Road in the south, adjoining Yinzhou District government, and Ningnan North Road to the west. Its total floor area is 21.09 ha. It consists of large-scale commercial shopping centre, five-star hotel, and high-rise urban appartment. These three function blocks consititute a grand urban building complex. Ningbo Yinzhou Wanda Plaza as the first project of the third generation of this commercial real estate, introduced interior pedestrian street for the first time, which has a milestone role in the development of Wanda commercial real estate. International shopping centre is situated in the northwest of the square, with a overall floorage of 260,000 m^2, which is the largest commercial plaza developed by Wanda so far. This plaza gathers various commercial forms, including a large-scale international chain supermarket, building materials supermarket, digital plaza, recreation plaza, cinema and food plaza, home furnishing plaza, and international department store. Moreover, it also collects 11 main stores, including Wal-mart, B&Q, HOLA, Xibaidu, Suning Appliance, Shipuda Hotel, Super Player Park, Super Star KTV, Wanda Cinemas, Yintai Dpartment Sore, Jishengweibang Home Furnishing, in addition with about 200 other well-known brand shops in China and abroad. Ningbo Yinzhou Wanda Plaza business circles have developed into the second biggest one in Ningbo.

哈尔滨香坊万达广场

HARBIN XIANGFANG WANDA PLAZA

2007 年开业

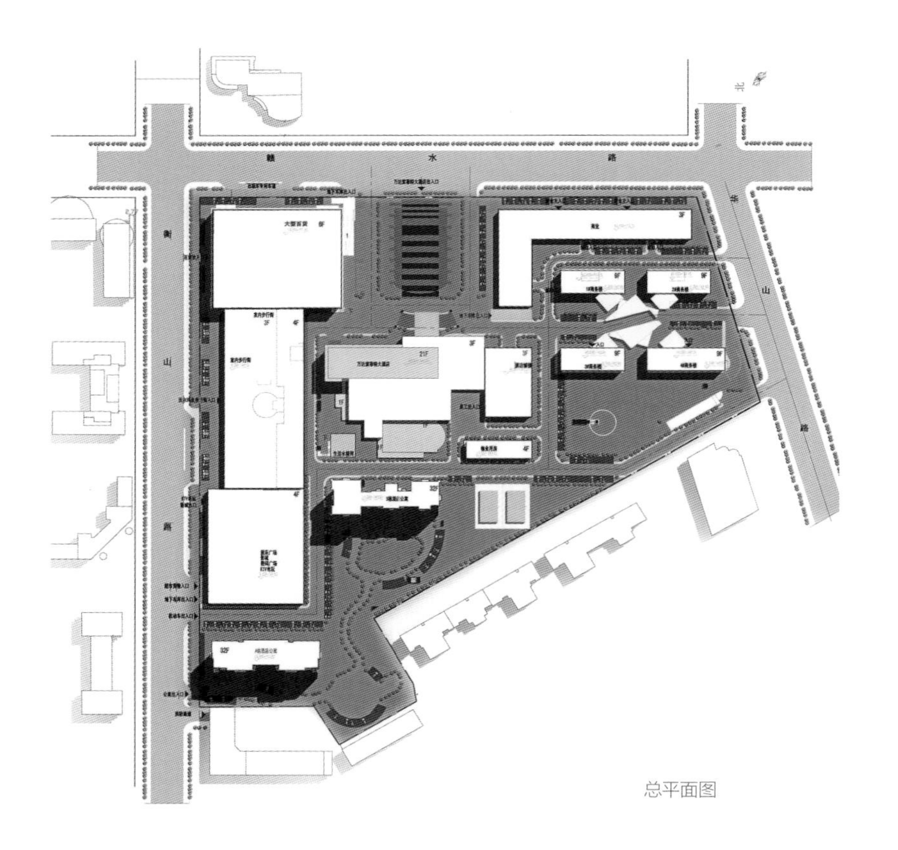

总平面图

广场日景

哈尔滨香坊万达广场坐落于哈尔滨市香坊区，2007 年 10 月 25 日开业，继中央大街、秋林商圈后，又一商业巨无霸迅猛崛起；比邻的哈尔滨万达索菲特大酒店，是哈市目前唯一一家白金五星级酒店。该区域现已成为哈尔滨市繁华的商业、商务中心区之一。

Harbin Xiangfang Wanda Plaza is located in Harbin City, Xiangfang District, opened in October 25, 2007, following the Central Avenue and Qiulin values, a rapid rise of commercial businesses. The Harbin Wanda Sofitel Grand Large Hotel is currently only a platinum five-star Hotel in Harbin city, which has become a central business district with hustling commerce.

项目总建筑面积 12 万平方米，现为哈尔滨首席城市综合体，拥有多元化的商业组合：万达百货、万达影城、大润发超市、国美电器、必胜客、大玩家超乐场、大歌星 KTV、高级餐饮等多种国际生活品质的完善配套及室内步行商业街，集购物、休闲、美食、娱乐于一体。新扩建完成的 IMAX 影厅，辐射区域的向心力，再次提升商圈的内在品质，绽放新的魅力。

The project has a total construction area of 120,000 m^2, and it is the chief city complex in Harbin, which has a plurality of business combination: Wanda Dept. Store, Wanda Cinemas, supermarket, GOME, Pizza Hut, Super Player Park, Super Star KTV, Senior Catering and other international quality of life to improve facilities and indoor pedestrian street, shopping mall, leisure, food, entertainment in one. The new extension of the IMAX hall radiates region of centripetal force, which enhances the intrinsic quality of bloom values and new charms.

成都锦华
万达广场

CHENGDU JINHUA WANDA PLAZA

2007 年开业

广场日景

成都锦华万达广场坐落于锦江区二环路东五段与锦华路交叉口,项目东邻锦华路,南依规划路,北靠二环路,西至老成仁路。

Chengdu Jinhua Wanda Plaza is located at the intersection of the east 5th segment of second ring road and Jinhua road. The site of the project is defined by the old Chengren road to the west, second ring road to the north, the planning road to the south and Jinhua rad to the east.

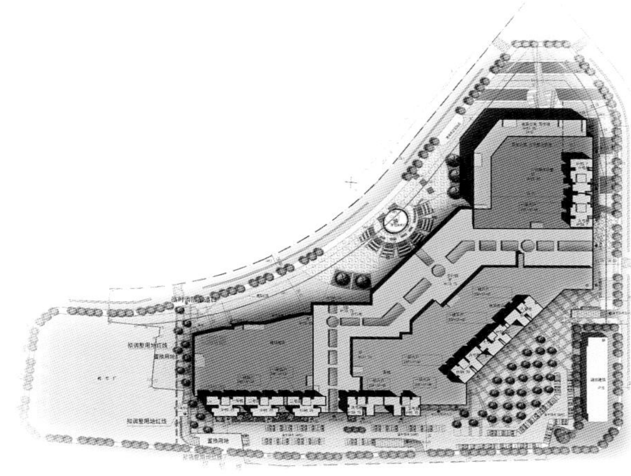

总平面图

成都锦华万达广场总建筑面积40万平方米，其中商业综合体7.8万平方米，囊括了万达百货、伊藤洋华堂、万达影城、大歌星KTV、大玩家超乐场、红杏酒家、国美电器等多家行业领航品牌，并由一条长355米，涵盖餐饮、休闲、时尚购物的室内步行街连接，覆盖现代城市生活各个领域，最大限度满足城市人群对商业设施与服务全方位、多元化的需求。

With total floor area of 400,000 m^2, of which 78,000 m^2 of commercial facilities, Chengdu Jinhua Wanda Plaza accommodates Wanda Dept. Store, Wanda Cinemas, Super Star KTV, Super Player Park, restaurants, GOME and international brands, and a 355 m shopping and entertainment street, providing service for all different aspects of metropolitan life to a great extent.

成都锦华万达广场一排30层现代高楼及业态组合丰富的购物中心为成都南片区注入活力，形成成都城市副中心，奠定了成都南区万达商圈地位。

With a row of 30 level high-end office buildings and mixed use shopping center, Chengdu Jinhua Wanda Plaza has introduced to south Chengdu with surprising vitality, and developed as a distinguished Commercial circle and sub center for this city.

装饰雕塑

影城入口

装饰雕塑

室内步行街中庭

室内步行街

西安李家村万达广场

XI'AN LIJIACUN WANDA PLAZA

2008 年开业

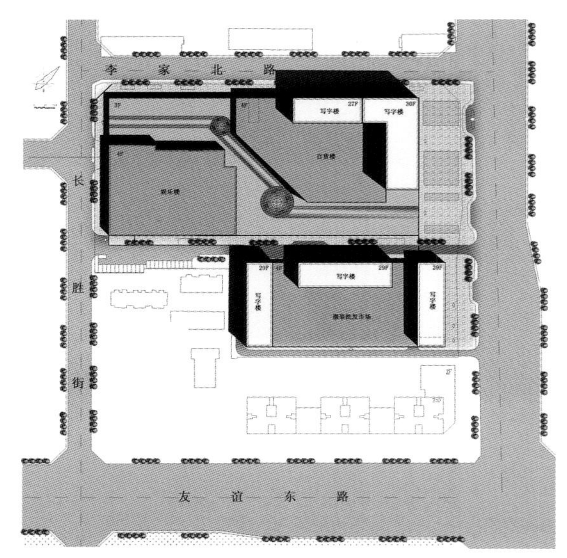

总平面图

西安李家村万达广场位于西安市碑林区雁塔路与友谊路交汇处，占地 72 亩，总建筑面积逾 35 万平方米，商业建筑面积 17 万平方米、国际公寓和写字楼 18 万平方米。引入了世界排名第一的零售业巨头沃尔玛超市、万达百货、国美电器、万达影城、大歌星 KTV、大玩家超乐场等各大主力店，在 3 万平方米的室内精品步行街中引进了诸多知名品牌。Z 字形、双中厅设计，完美的动线，有机地衔接各业态，客流汇聚，绝无被遗忘的角落。阳光中庭、观光电梯、自动扶梯、专设货梯全力保障。国际商业享受，全新都市体验，尽在万达广场！

Xi'an Lijiacun Wanda Plaza is situated at the intersection of Yanta Road and Youyi Road of Beilin District, Xi'an. The plaza covers an area of 48,000 m^2, with a gross floor area of more than 350,000 m^2, of which 170,000 m^2 is the retail part and 180,000 m^2 is taken up by apartment and office buildings. The anchor stores of the plaza include world No.1 retailer corporation – Wal-Mart, Wanda Dept. Store, GOME, Wanda Cinemas, Super Star KTV, Super Player Park, The 30,000 m^2 indoor boutique shopping street has also introduced many big brands. The Z-shaped circulation, double atrium and automatic escalator link all the different programs into an organic whole and gives convenient access to every single corner of the mall. Sun-bathed atriums, automatic lifts and escalators, exclusive service lifts can all be found in the plaza which brings a brand new world-class retail experience to the city!

广场日景

北京石景山万达广场

BEIJING SHIJINGSHAN WANDA PLAZA

2008 年开业

北京石景山万达广场为北京市石景山区鲁谷银河商务区商业商务地标项目，位于北京市石景山区石景山路地铁八宝山站西侧，北临石景山路，南接银河中街，东临鲁谷大街，西接银河东街，总建设用地面积约 5.2 万平方米；总建筑面积约 28 万平方米；由 6 座写字楼、4 层商业裙房和 1 座五星级酒店组成，涵盖办公、餐饮、娱乐、休闲、购物、影城、KTV 和酒店等业态。地上分东区和西区，东区由 1 座五星级酒店和 2 座写字楼组成，西区由 4 座写字楼和 4 层商业裙房组成；地下 2 层，设有人防和停车场。

Beijing Shijingshan Wanda Plaza is a landmark project in Yinhe CBD, Lugu sub-district, Shijingshan District, Beijing. The project lies next to Babaoshan station of metro line on its west side, bordering shijingshan Road on its north side, Middle Yinhe Street on its south side, Lugu Avenue on its east side and East Yinhe on its west side, with a total area of 52,000 m^2, and a gross floor area of 280,000 m^2. The plaza include 6 office towers, 4 retail podiums and a five-star hotel, encompassing a wide range of programs, such as office, F&B, entertainment, leisure, shopping, cinema, KTV and hotel. The above-grade part consists of an east section and a west section. The east section includes one five-star hotel and 2 office towers and the west section includes four office towers and four retail podiums. The basement part has 2 floors, including civil air defense and car park.

主广场入口

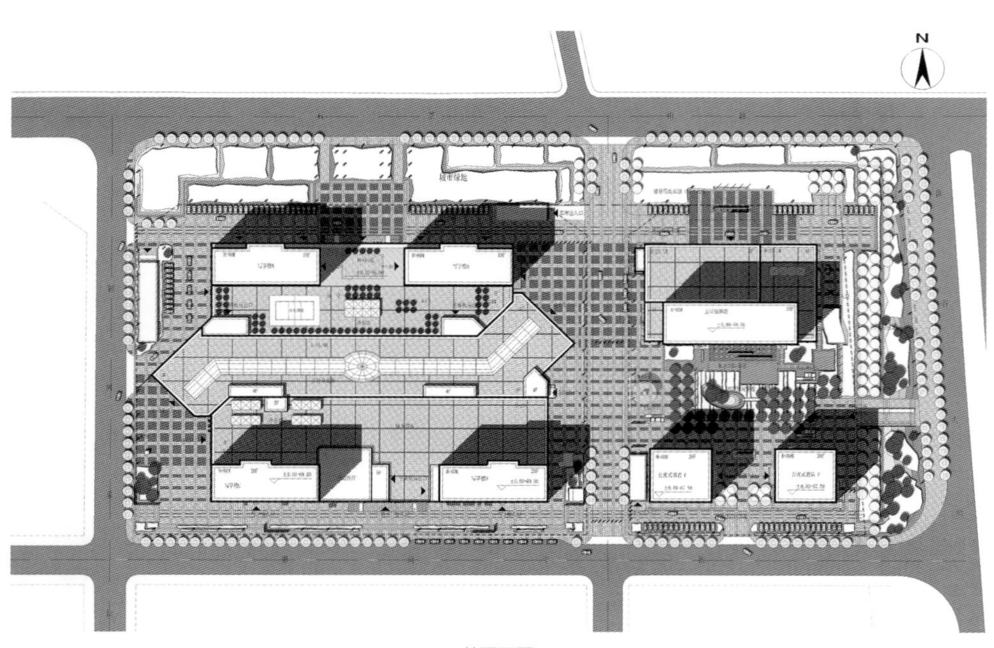

总平面图

万达广场
WANDA PLAZA
国美电器
UNIQLO
万达广场
室内步行街
PEDESTRIAN STREET
万千百货
VAN'S
运动100
SPORT 100
国美电器
GOME
家乐福
CARREFOUR
大玩家超乐场
S-PLAYER
大歌星KTV
SUPER STAR KTV
万达国际影城
WANDA CINEMAS
万达铂尔曼大饭店
PULLMAN
写字楼A·B座
TOWER A·B
写字楼C、D座
TOWER C、D
公寓
APARTMENT

室内步行街

万千百货
VAN'S DEPT. STORE
ZARA

室内步行街

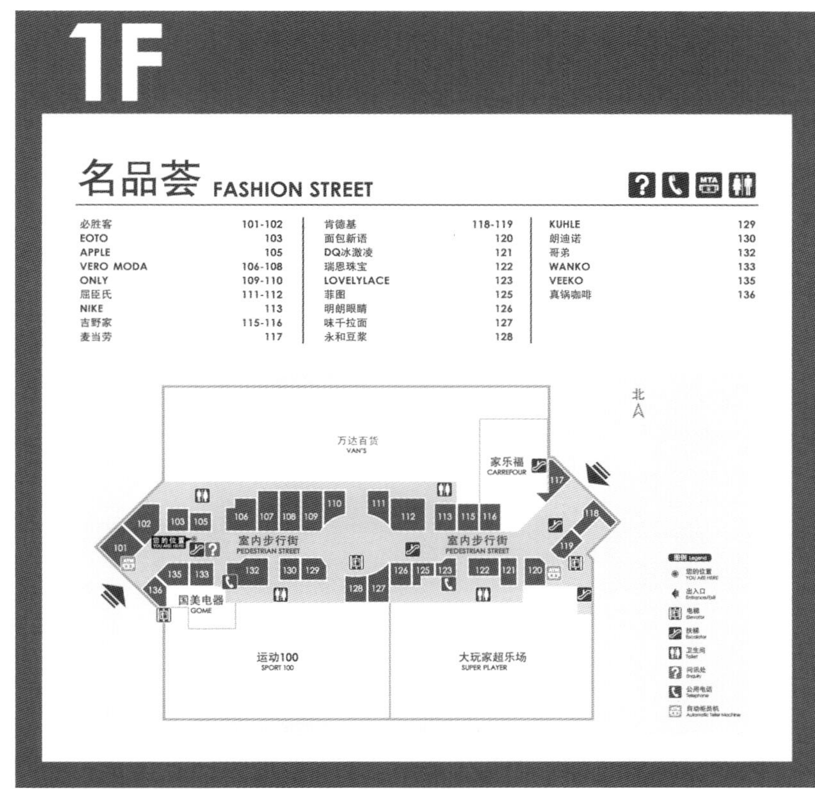
1F
名品荟 FASHION STREET
必胜客 101-102
EOTO 103
APPLE 105
VERO MODA 106-108
ONLY 109-110
屈臣氏 111-112
NIKE 113
吉野家 115-116
麦当劳 117
肯德基 118-119
面包新语 120
DQ冰激凌 121
瑞恩珠宝 122
LOVELYLACE 123
菲图 125
明朗眼睛 126
味千拉面 127
永和豆浆 128
KUHLE 129
朗迪诺 130
哥弟 132
WANKO 133
VEEKO 135
真锅咖啡 136
北
万达百货
VAN'S
家乐福
CARREFOUR
室内步行街
PEDESTRIAN STREET
国美电器
GOME
运动100
SPORT 100
大玩家超乐场
SUPER PLAYER

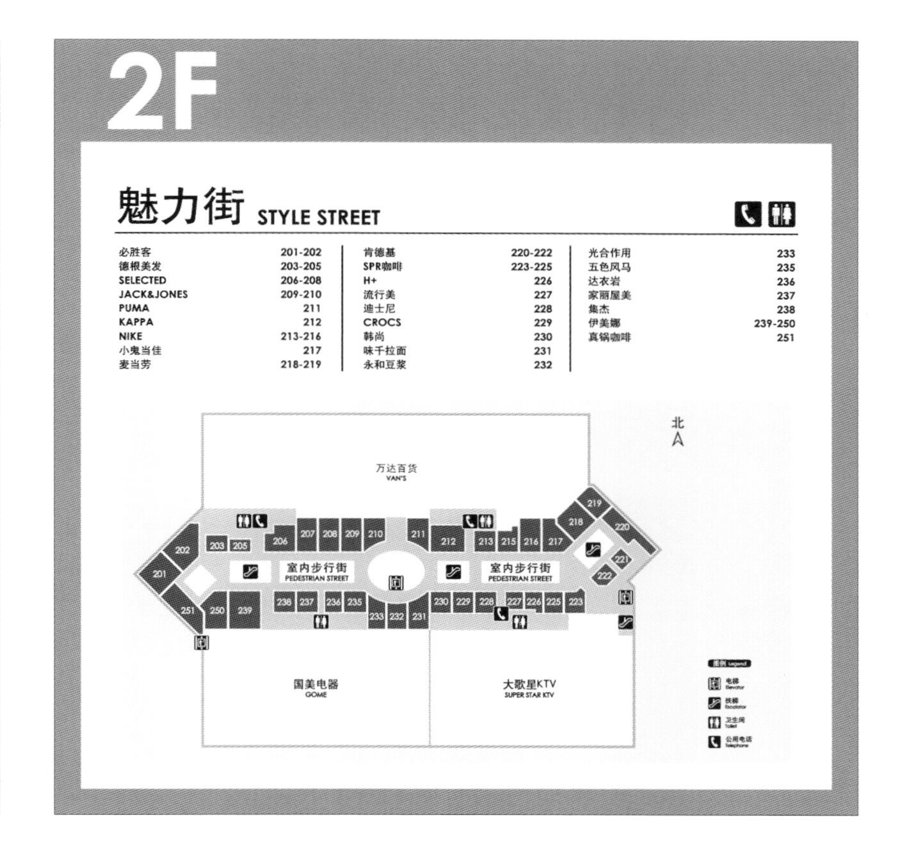
2F
魅力街 STYLE STREET
必胜客 201-202
德根美发 203-205
SELECTED 206-208
JACK&JONES 209-210
PUMA 211
KAPPA 212
NIKE 213-216
小鬼当佳 217
麦当劳 218-219
肯德基 220-222
SPR咖啡 223-225
H+ 226
流行美 227
迪士尼 228
CROCS 229
韩尚 230
味千拉面 231
永和豆浆 232
光合作用 233
五色风马 235
达衣岩 236
家丽屋美 237
集杰 238
伊美娜 239-250
真锅咖啡 251
北
万达百货
VAN'S
室内步行街
PEDESTRIAN STREET
国美电器
GOME
大歌星KTV
SUPER STAR KTV

品牌落位图

成都万达
索菲特大饭店
SOFITEL WANDA CHENGDU

2003 年开业

成都万达索菲特大饭店坐落于美丽的锦江河畔，位于一环内城最繁华地段锦江区滨河中路，是大连万达集团在中国西部投资修建的第一家饭店，也是法国雅高集团在中国西部地区经营管理的第一家饭店，是首批国家"金叶"级绿色旅游饭店。

Sofitel Wanda Chengdu is located in the beautiful Jinjiang waterfront, which is also on the Binhe Middle Road of Jinjiang District in the most flourishing area of the First Ring inner city. It is the first restaurant in western China that is invested and built by Dalian Wanda Group. It is also the first restaurant in westen China operated and managed by France Accor Group, as well as the first batch of national "Jinye"-class green tourism restaurants.

酒店占地 0.68 万平方米、总建筑面积 3.9 万平方米，共 23 层，拥有 262 间精致时尚豪华客房及套房和 11 间规模不同的会议室。精致典雅的设计，体贴周全的服务，处处尽显法式浪漫，让宾客充分享受与众不同的尊贵体验。总体设计顺应地形，灵活的体量组合，沿江弧形展开的裙房及塔楼客房最大限度的纳入锦江美景。

The hotel occupies 6,800 m^2 for the total floor area. Its overall floorage is 39,000 m^2, with 23 storeys. It owns 262 delicate, fasionable and luxury guest rooms and suites, and 11 meeting rooms in diferent scales. Its delicate design, considerate service, with French romance everywhere, provide guests an unique and luxury experience. The general design adjusts the geography, with flexible size groups, arc-shaped group buildings and tower buildings along the river, which all absorb the beautiful scene of Jinjiang maximumly.

酒店日景

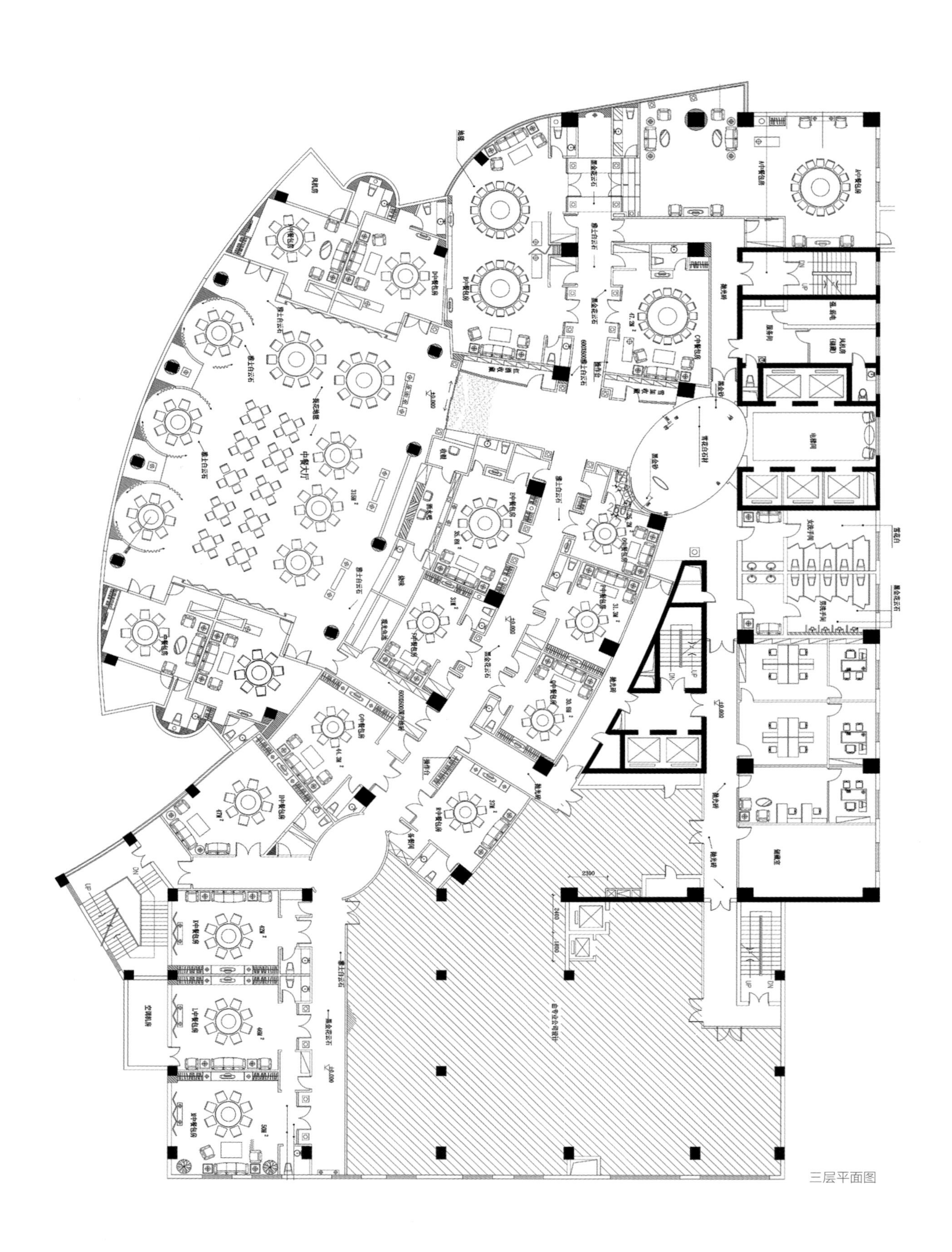

三层平面图

酒店大堂

酒店夜景

北京万达
索菲特大饭店
SOFITEL WANDA BEIJING

2006 年开业

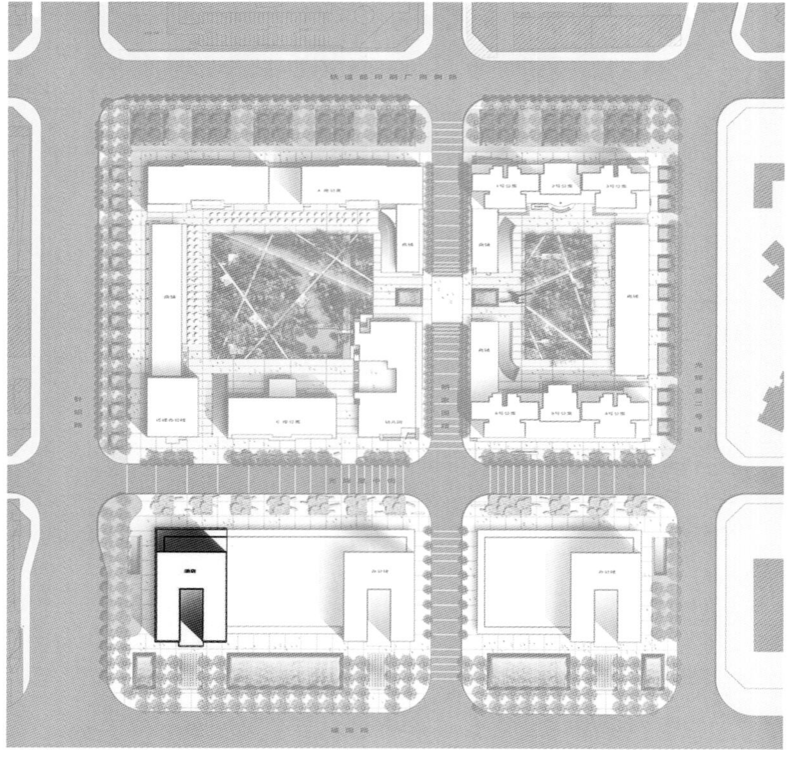

总平面图

北京万达索菲特大饭店位于北京市中央商务区核心地段，紧邻东三环和国贸中心，是北京 CBD 万达广场建筑群的重要组成部分。

Located in the core CBD of Beijing, next to East 3rd Ring Road and China World Trade Center, Sofitel Wanda Beijing forms an important part to Beijing CBD Wanda Plaza.

北京万达索菲特大饭店面积约 4.8 万平方米，是法国索菲特品牌酒店在亚太区的旗帜酒店。酒店高 27 层，内设 417 间标准客房与套房，房间豪华舒适。同时拥有面积不等的 10 间多功能厅，其中大宴会厅面积将近 1400 平方米。饭店设有餐厅、葡萄酒及雪茄吧、大堂酒廊、行政酒廊、健身中心及蓝蔻 SPA 等，为商务下榻、会议洽谈及休闲度假的理想之选。

As the French Sofitel Hotel's flagship hotel in the Pacific Asia Region, the 27-story Beijing Wanda Sofitel Hotel occupies a gross floor area of 48,000 m^2, offering 417 luxury and comfortable guest rooms and suites, 10 multi-function rooms of different sizes, and a 1,400 m^2 ballroom. The hotel also offers restaurants, wine and cigar bar, lobby lounge, executive lounge, gym and Lancome SPA, making it an ideal choice for your stay, conference and holiday.

酒店日景

酒店大堂

总统套房卧室

室内游泳池

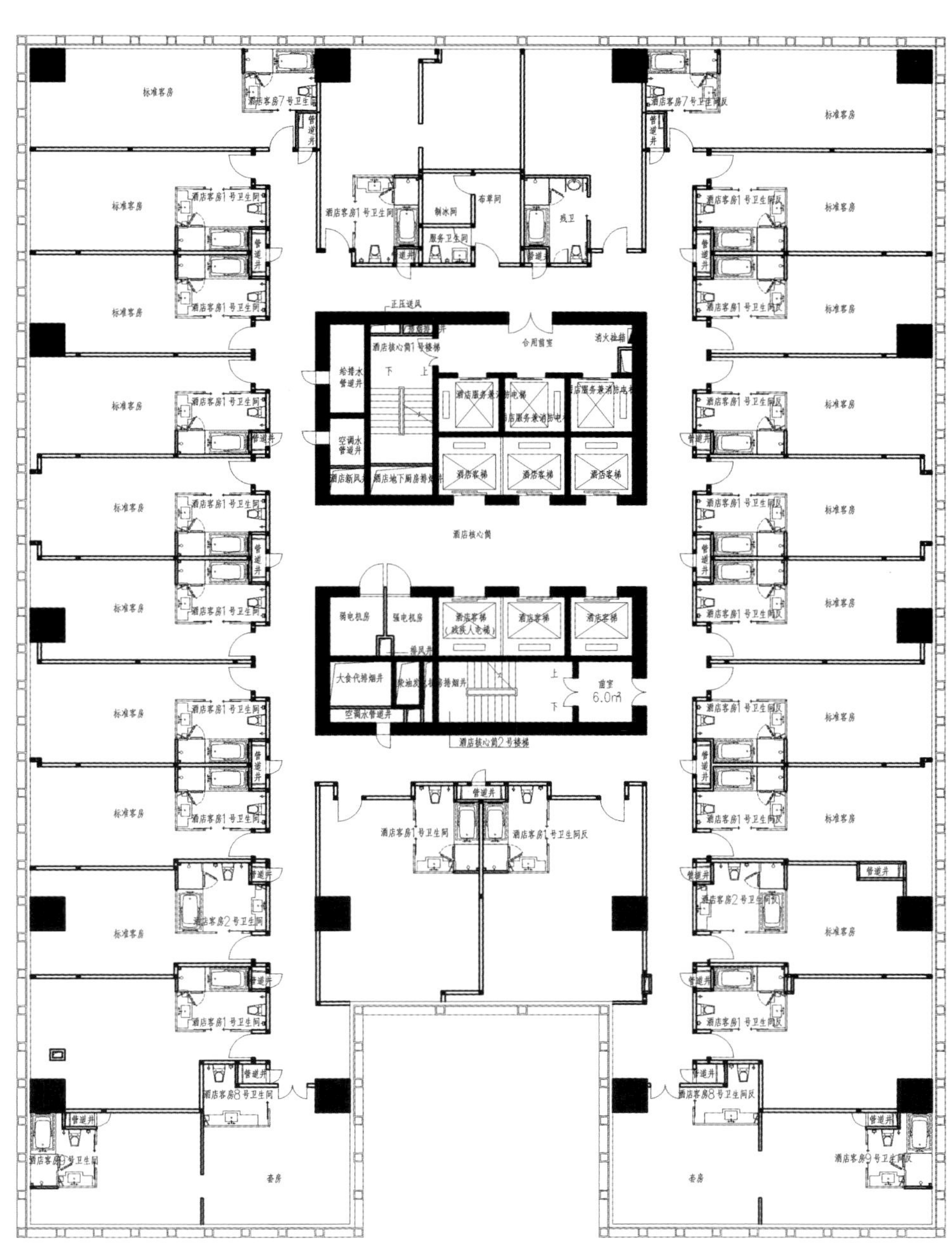

标准层平面图

哈尔滨万达索菲特大酒店
SOFITEL WANDA HARBIN

2007 年开业

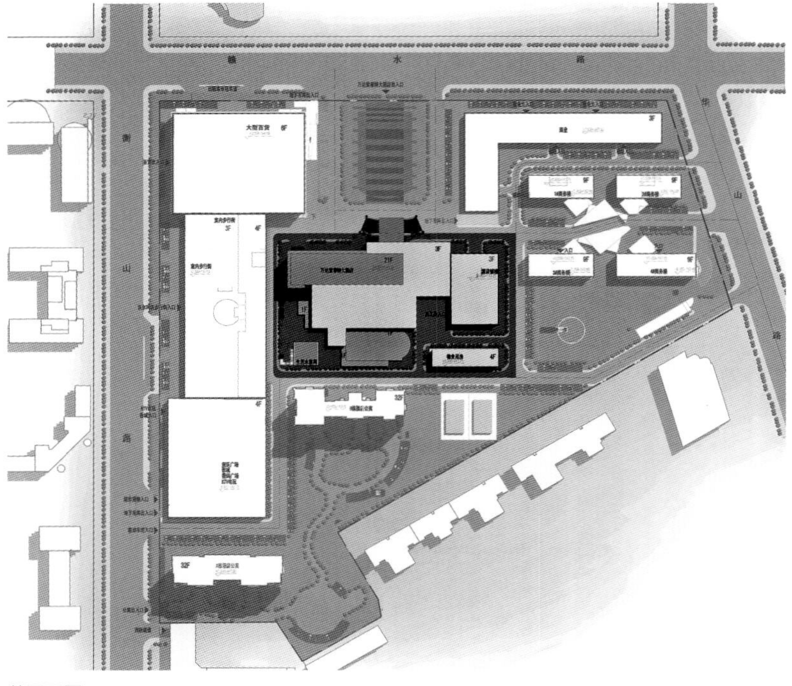

总平面图

哈尔滨万达索菲特大酒店是哈尔滨市目前唯一一家白金五星级酒店。大酒店是由大连万达集团投资，法国雅高管理集团管理的一家国际豪华五星级品牌酒店。酒店内完备的配套设施，为企业商务接待、会议洽谈、商业展示与推介等提供殿堂级服务理念，地道法式风情的酒店风格与服务特色。酒店集中、法两国的优雅，和谐于一身，树立哈尔滨市现有的五星级酒店新的行业标准。

Sofitel Wanda Harbin is currently one of few platinum five-star hotesl in Harbin. It is an international luxury five-star hotel invested by Dalian Wanda Group and managed by Accor Hotels & Resorts Management French. It has comprehensive facilities, which provides authentic luxury French-style service and concept for the business receptions, conference, commercial exhibition and recommendation. The hotel brings together elegance and harmony from France and China in one, and also shapes the new service concept and industry standards for the existing five-star hotels in Harbin.

酒店位于哈尔滨市香坊区赣水路与衡山路的十字路口，拥有 322 间客房及独立的索菲特会所，三间风格各异的餐厅，包括亚洲特色美食和传统法式大餐餐厅、日式餐厅及中餐厅，2 个酒吧，600 平方米的多功能宴会厅和多种会议室，SPA 会馆、健身中心和室内泳池等休闲娱乐场所。

The hotel located in 68 Ganshui Road, Xiangfang District, Harbin, has 322 rooms and independent Sofitel Club. It has 3 restaurants including Asian characteristic food and traditional French restaurant, Japanese restaurant and Chinese restaurant; 2 bars; 600 m^2 multi-functional banquet hall and a variety of conference rooms; Spa; fitness centre; indoor swimming pool and other entertainment venues.

酒店夜景

酒店大堂

豪华套房

餐厅

豪华套房

餐厅

会议室

酒店大堂

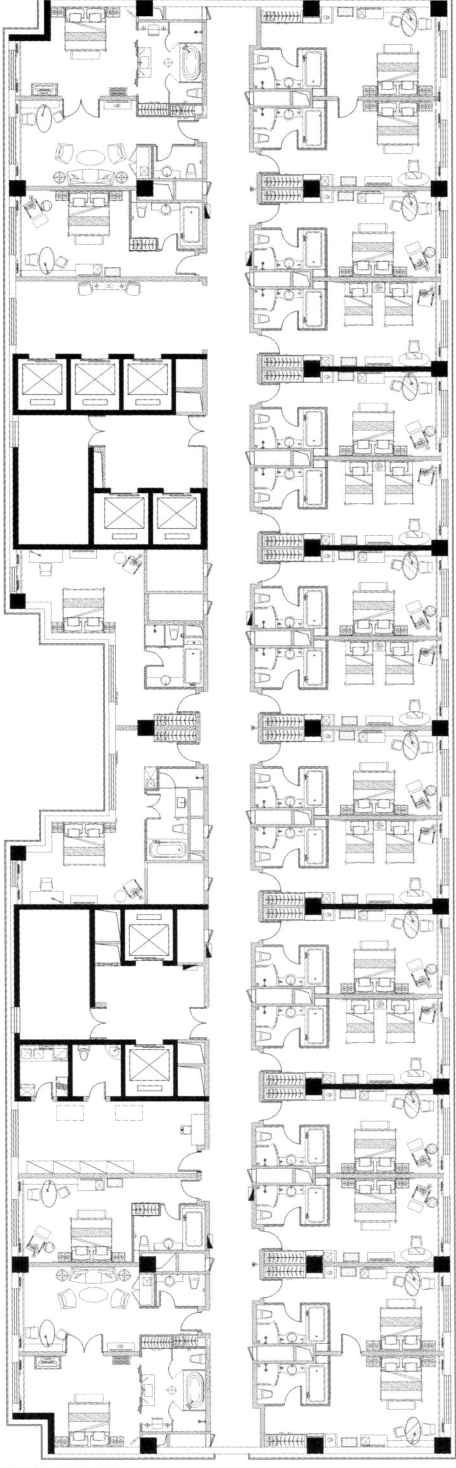

标准层平面图

餐厅

大堂吧

宁波万达索菲特大饭店

SOFITEL WANDA NINGBO

2008 年开业

酒店外立面日景

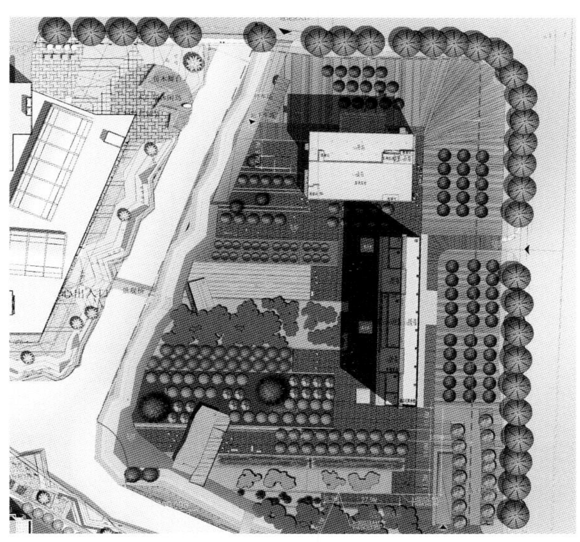

总平面图

宁波万达索菲特大饭店地处鄞州新区，毗邻大型购物中心万达广场，与都市商业区的繁华和现代相得益彰。酒店楼高 19 层。这座国际五星级酒店的风格吸纳了法兰西的浪漫情怀和中国的古典文雅，并与甬城深厚的旅游文化相融合。

宁波万达索菲特大饭店共拥有 290 间豪华客房和套房。酒店设有 3 间风格现代、设施齐全的会议厅，另有 6 间风味独特的餐厅。精心打造的乐吧极具爵士韵味。酒店另有 1 间大型豪华宴会厅，可服务于各类婚庆宴请、小型私人派对或大型鸡尾酒会。酒店还拥有室内恒温游泳池和专业 SPA 服务，为追求健康和休闲的宾客提供完美的体验。

Sofitel Wanda Ningbo is located in Yinzhou New District, closed to grand shopping centre Wanda Plaza. It complements with the flouring and modern metropolitan commercial area. It has a total height of 19 floors. The character of this international five-star hotel absorbed French romantic feeling and Chinese classic elegance, also combined with the deep tourism culture of Ningbo City.

The hotel has a total of 290 luxurious guest rooms and suites. The hotel is equipped with 3 modern-style conference rooms with complete facilities, 6 unique style restaurants and a elaborately-built Jazz Music Bar. It also owns a large luxury banquet hall, which could be used for all kinds of wedding, small private party and big cocktail party. In addition, there is an indoor thermostatical swimming pool and specialized SPA service, which provides a perfect experience for the guests who pursue fitness and leisure.

酒店夜景

酒店夜景

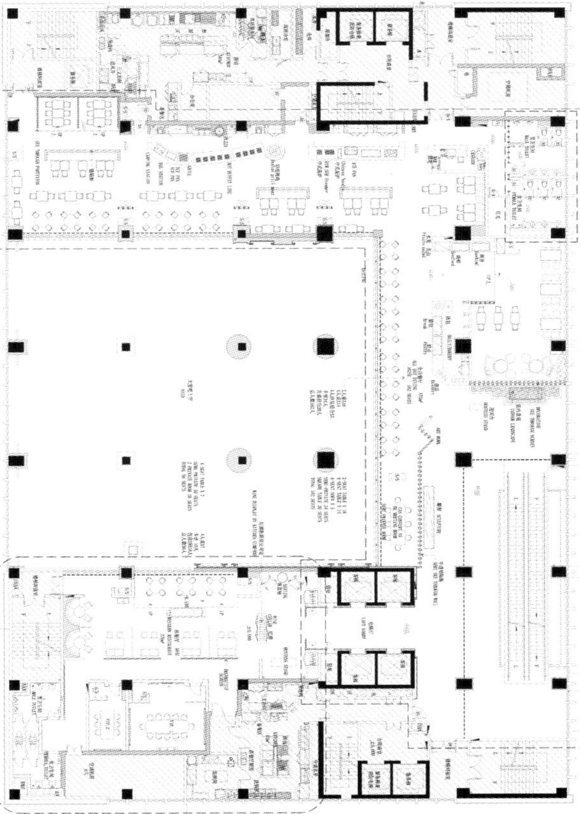
局部平面图

北京万达
铂尔曼大饭店
PULLMAN BEIJING WEST WANDA

2008 年开业

酒店日景

酒店夜景

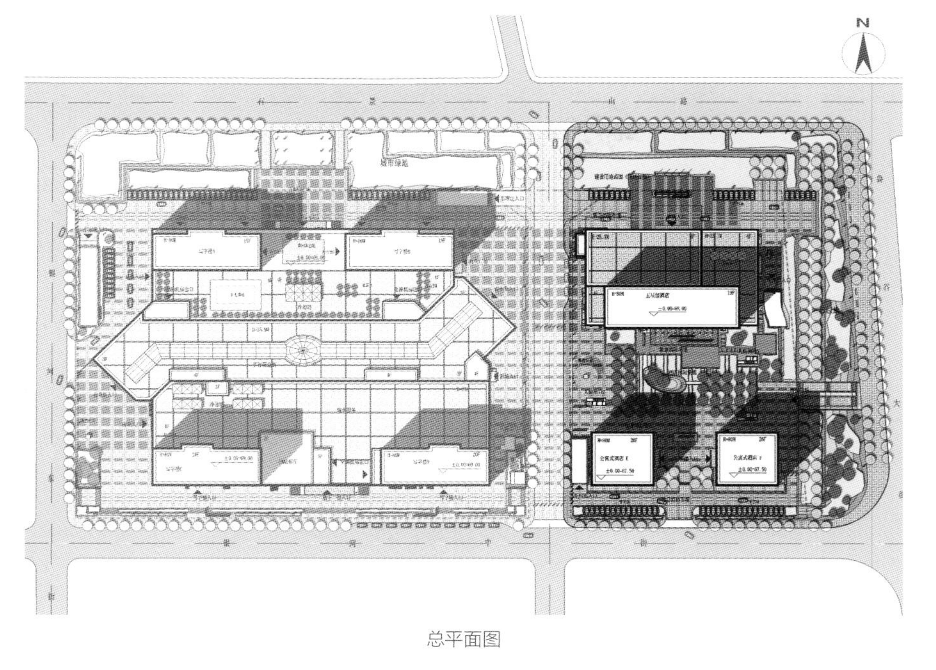

总平面图

北京万达铂尔曼大饭店位于西长安街的商业中心——石景山万达广场内，地理位置优越，公共交通发达。饭店为客人提供了量身定做的个性化高档商务住宿和轻松舒适的环境。

Pullman Beijing West Wanda enjoys a premium location in Shijingshan Wanda Plaza, West Chang'an Avenue CBD with convenient access to public transport. The hotel offers personalized high-end accommodation and pleasant and comfortable environment.

北京万达铂尔曼大饭店总建筑面积约 3.58 万平方米。酒店高 19 层，内设有 312 间标准客房与套房，房间豪华舒适。同时拥有面积不等的多功能厅，其中大宴会厅面积将近 1400 平方米，配有高科技沟通系统和先进的会议服务设施。

The 19-story Pullman Hotel occupies a gross floor area of 35,800 m^2, offering 312 luxury and comfortable guest rooms and suites, multi-function room of different sizes and a 1,400 m^2 ballroom, equipped with state-of-the-art communication system and conference facilities.

饭店设有餐厅、大堂酒廊、行政酒廊、健身中心及各式水疗 SPA 等，为商务下榻、会议洽谈的理想之选。

The hotel also offers restaurants, lobby lounges, executive lounges, gym and a variety of SPAs, making it the best choice for stay, conference and business meetings.

酒店目景

酒店大堂

酒店豪华套房

酒店大堂局部

中华大宴会厅

PART 5 项目索引 INDEX OF THE PROJECTS

项目索引 2009 及 2009 前
INDEX OF THE PROJECTS IN 2009 & BEFORE 2009

万达广场/ WANDA PLAZAS

洛阳万达广场
LUOYANG WANDA PLAZA
2009.12

西安民乐园万达广场
XI'AN MINLE PARK WANDA PLAZA
2009.12

南京建邺万达广场
NANJING JIANYE WANDA PLAZA
2009.12

重庆南坪万达广场
CHONGQING NANPING WANDA PLAZA
2009.12

青岛 CBD 万达广场
QINGDAO CBD WANDA PLAZA
2009.11

沈阳太原街万达广场
SHENYANG TAIYUANJIE WANDA PLAZA
2009.11

上海周浦万达广场
SHANGHAI ZHOUPU WANDA PLAZA
2009.09

苏州平江万达广场
SUZHOU PINGJIANG WANDA PLAZA
2009.09

北京石景山万达广场
BEIJING SHIJINGSHAN WANDA PLAZA
2008.12

西安李家村万达广场
XI'AN LIJIACUN WANDA PLAZA
2008.05

成都锦华万达广场
CHENGDU JINHUA WANDA PLAZA
2007.12

哈尔滨香坊万达广场
HARBIN XIANGFANG WANDA PLAZA
2007.10

北京 CBD 万达广场
BEIJING CBD WANDA PLAZA
2006.12

上海五角场万达广场
SHANGHAI WUJIAOCHANG WANDA PLAZA
2006.12

宁波鄞州万达广场
NINGBO YINZHOU WANDA PLAZA
2006.12

南昌八一万达广场
NANCHANG BAYI WANDA PLAZA
2003.08

酒店/ HOTELS

重庆万达艾美酒店
LE MERIDIEN CHONGQING, NAN'AN
2009.12

青岛万达艾美酒店
LE MERIDIEN QINGDAO
2009.11

北京万达铂尔曼大酒店
PULLMAN BEIJING WEST WANDA
2008.12

宁波万达索菲特大饭店
SOFITEL WANDA NINGBO
2008.12

哈尔滨万达索菲特大酒店
SOFITEL WANDA HARBIN
2007.10

北京万达索菲特酒店
SOFITEL WANDA BEIJING
2006.12

成都万达索菲特大饭店
SOFITEL WANDA CHENGDU
2003.12

大连万达商业地产股份有限公司
DALIAN WANDA COMMERCIAL ESTATE CO., LTD.

大连万达商业地产股份有限公司成立于2002年9月，2009年12月整体变更为股份有限公司，注册资本37.36亿元人民币，是大连万达集团旗下商业地产投资及运营的唯一业务平台。公司的主营业务为商业地产投资及运营管理，核心产品是以“万达广场”命名的万达城市综合体。

The Dalian Wanda Commercial Estate Co., Ltd. was established in September 2002 and subsequently converted as a whole into a joint-stock company in December 2009. With a registered capital of RMB 3.736 billion, it is Dalian Wanda Group's only business platform for commercial property investments and operations. The company's core business includes commercial property investments and operations management, and its core product is the urban complex named as "Wanda Plaza".

截至2011年底，本公司已经在全国31个省、自治区和直辖市的61个重点城市投资建设了87个万达广场，其中已开业49个万达广场，正在运营26家五星级或超五星级酒店。本公司计划在2012年新开业17个万达广场，新运营8家五星级或超五星级酒店。作为率先实现全国布局的大型商业地产投资及运营商，公司在中国商业地产行业内居于绝对领先地位，是中国商业地产的领军企业。

By the end of 2011, the company had invested in 87 Wanda Plazas in 61 key cities across 31 provinces, autonomous regions and municipalities in China. Currently, 49 Wanda Plazas and 26 5-star or super 5-star hotels are in operation. The company plans to open 17 more Wanda Plazas and operate 8 more 5-star or super 5-star hotels in 2012. As the first large-scale commercial property investor and operator to achieve nationwide distribution, the company occupies the absolute leadership in China's commercial real estate industry and is the pioneer of China's commercial real estate.

目前本公司主营业务主要包括以大型购物中心为主体的商业中心投资与运营（简称“商业中心”）；五星级及超五星级酒店的开发与运营（简称“高级酒店”）；商业运营及物业管理（简称“商业管理”）；写字楼、公寓和住宅的开发销售（简称“销售物业”）在内的四大核心业务板块，公司四大业务板块之间相互支持、相互提升，形成了一个有机业务整体。这种以商业中心为核心、充分发掘业务板块之间联动效应的城市综合体开发与运营管理，构成了本公司的核心业务发展模式。

Currently, the company's core business includes: investment and operation of commercial centers which are predominantly large shopping centers (referred to as "commercial centers"); development and operation of 5-star or super 5-star hotels (referred to as "luxury hotels"); business operation and property management (referred to as "commercial management"); development and sales of office, apartment and residential buildings (referred to as "property sales"). These core business segments provide mutual support to one another and help each other rise to higher levels, thereby forming an organically integrated business structure. The company's core business model is to develop and operate the urban complexes with commercial centers as their core while fully exploring the linkage effects among different business segments.

以“万达广场”命名的城市综合体是目前国内外领先的商业地产产品，具有显著的社会经济效益：为商贸、文化、娱乐、体育、餐饮等第三产业提供广阔的发展平台，从而带动所在城市的产业结构调整；全方位满足和创造新的消费需求，从而有效拉动和刺激消费；打造新的城市中心，完善城市区域功能，促进城市均衡发展；创造大量的就业岗位；汇聚了众多国内外知名企业，实现商业持续繁荣，创造持续巨额税源。

The urban complex, named "Wanda Plaza", is currently the leading commercial property product at home and abroad, with significant social and economic benefits, such as: providing a broad development platform for service industries such as commerce & trade, culture, entertainment, sports, and food & beverages, thereby promoting industrial restructuring in the cities where the complexes are located; fully satisfying and create consumer demands, thus effectively driving and stimulating consumption; create new urban commercial centers, improving the function of urban areas, and promoting the balanced development of the cities; creating a large number of jobs; Bringing together dozens of well-known enterprises at home and abroad, sustaining business prosperity and creating continued large tax revenue sources.

本公司在市场竞争中逐渐形成独有的核心竞争优势，主要体现在：完整的产业链、突出的资源获取能力、独特的“订单地产”模式、成熟的盈利模式、高效的管控及执行能力、优秀的企业文化和卓越的品牌影响力等方面。凭借上述核心竞争力，本公司在全国范围内实现了万达城市综合体产品模式的快速复制，成为目前国内唯一实现全国布局的城市综合体投资及运营企业。

The company has gradually formed its unique core competitive advantages which are reflected in the integrated business chain, outstanding ability in procuring resources, unique "customized developing model", mature profit model, efficient control and enforcement capacity, excellent corporate culture and preeminent brand influence. Relying on these core competencies, the company has rapidly replicated the product model of Wanda Plazas nationwide, thereby becoming the only urban complex investor and operator to achieve national distribution currently.

未来，本公司将利用“万达广场”的品牌影响力，继续重点拓展一线城市市场，适当扩大在二、三线城市的布局，并力争在各区域市场占有一定的市场份额，从而巩固并进一步扩大全国性布局的长期竞争优势。

Going forward, the company will utilize the brand influence of "Wanda Plaza" to continue to focus on expanding the market in first-tier cities, strategically expand its distribution layout in second and third tier cities and strive to take hold of a certain amount of market share in each regional market, hence consolidating and further enhancing the long-term competitive advantages of its national distribution layout.

本公司的长期战略目标是：巩固亚洲商业地产排名第一的领先优势，力争成为全球商业地产行业的领军企业，最终成为全球持有物业面积最大的商业地产企业，从而实现“国际万达，百年企业”的愿景。

The company's long-term strategic goal is to consolidate its status of being Asia's No.1 commercial property company, strive to become the leading developer and operator in the global commercial property industry, and eventually become the largest global commercial property company in terms of the GFA of investment property, thereby achieving its vision of "International Wanda and Century-Old Enterprise".

万达商业规划研究院有限公司

WANDA COMMERCIAL PLANNING & RESEARCH INSTITUTE CO., LTD.

万达商业规划研究院有限公司注册成立于2007年，是全国唯一一家从事商业及文化旅游项目规划设计，同时进行全过程管控的技术管理和研究机构。万达商业规划研究院擅长商业业态规划、文化旅游项目规划及大型购物中心、五星级酒店、秀场及演艺剧场、主题公园等大型公建项目的建筑设计，是万达集团的技术归口管理部门。

Wanda Commercial Planning & Research Institute Co., Ltd. was established in 2007, as China's only technical management and research institution engaged in the planning and designing of commercial and cultural tourism projects at the same time of conducting whole-course control. The company is proficient in the planning of commercial activities, the planning of cultural tourism projects as well as the architectural designing of large-scale shopping centers, 5-star hotels, show theatres, performing arts theatres, theme parks and other public building projects. It is Wanda Group's competent department for technical management.

万达商业规划研究院下设文化旅游分院，是万达集团文化旅游项目的设计、管控及研发部门。文化旅游项目涵盖高端酒店群、主题公园、秀场、高尔夫球场、影视基地、滑雪场、旅游小镇等诸多业态，致力于打造国际一流的文化旅游度假胜地。

One part of the Institute is its Cultural Tourism Branch, which handles the designing, controlling and R&D of Wanda Group's cultural tourism projects, including high-end hotel groups, theme parks, show theatres, golf courses, movie & TV bases, ski resorts, tourism towns and other numerous businesses. The Branch is dedicated to crafting international top-level cultural tourism resorts.

万达商业规划研究院现有员工近300人，均为兼具建筑设计及房地产公司商业管理经验的复合型人才。其中，拥有各专业注册执业资格者大于员工总数的40%，在全国大型设计机构中比例最高；拥有高级技术职称者占员工总数的25%；研究生以上学历者占员工总数的26%。万达商业规划研究院是万达核心竞争力之一。

The Institute currently has nearly 300 employees on its staff, all versatile professionals with experience in both architectural designing and commercial management in real estate companies. Among them, over 40% have registered practitioners of various specialties, which is highest among all large designing institutions in China. 25% of them have senior technical titles and 26% of them hold Master's degrees and above. Wanda Commercial Planning & Research Institute is one of Wanda Group's core competitive edges.

万达商业规划研究院将秉承“求实、求是、求精；安全、品质、节能”的理念，不断提升管理和技术水平，力争成为商业及文化旅游规划设计领域内具有国际影响力的专业机构。

Wanda Commercial Planning & Research Institute will adhere to the concepts of "seeking facts, truth and refinement" and "guaranteeing safety, quality and energy conservation" and ceaselessly upgrade its management and technical capabilities, striving to become a specialized institution with international influence in the field of the planning and designing of commercial and cultural tourism projects.

2009万达商业规划研究院

WANDA COMMERCIAL PLANNING
& RESEARCH INSTITUTE 2009

郭薇　万理　刘笑宇
唐海江　徐涛
修明　吕彬锋
霍小虎
李晶

侯卫华　王晖　黄大卫　文善平　雷磊
兰峻文　王魏巍　梁阔　刘悦飞
赖建燕　曾少卿　刘婷　童球　臧久龙
张振宇　熊伟　孙光磊
孙培宇　王元　李甜　王明妍
谷建芳　张帆　李彬

王燕

李琛　杨彬

王弘成

吴绿野　王群华

王鑫　朱莹洁　黄勇

刘晓　尚海燕　叶宇峰

田杰　刘婷　梁爽　李楠

李峻　邹成江　张振宇　马红

李成斌　郭双　孙多斌　鲍力　李杨

曹亚星　魏成刚　张琳

范珑　田江　朱其玮　李建伟

李峥　杨世杰

夏洪兴

杨旭　曹冰　潘朝晖

倪群　林胜刚　苗凯峰

杨纪元　王绍合　于瑞勇　刘平

张帅克　严铁钰　杨滨　谷全

任小兵　吴晓璐　张涛　莫力生

胡伟　徐俊　于光炤　佟嘉

杨根朝　赵新宇　秦好刚　刘佳

龙向东　李文娟　田迎斌

常宇　徐小莉

公寓
APARTMENT
写字楼C·D座
TOWER C·D
写字楼A、B座
TOWER A、B

北京石景山万达广场

SOFITEL
万达国际电影城
WANDA INTERNATIONAL CINEMAS

北京CBD万达广场

图书在版编目（CIP）数据

万达商业规划2009 / 万达商业规划研究院主编.
—北京：中国建筑工业出版社，2013.5
ISBN 978-7-112-15370-1
Ⅰ. ①万… Ⅱ. ①万… Ⅲ. ①商业区—城市规划—中国
Ⅳ. ①TU984.13
中国版本图书馆CIP数据核字(2013)第078749号

责任编辑：徐晓飞 张 明 徐 冉 李 鸽
美术编辑：康 宇
装帧设计：洲联集团·五合国际·五合视觉
责任校对：姜小莲 王雪竹

万达商业规划2009
万达商业规划研究院 主编
*
中国建筑工业出版社出版、发行（北京西郊百万庄）
各地新华书店、建筑书店经销
北京雅昌彩色印刷有限公司制版
北京雅昌彩色印刷有限公司印刷
*
开本：787×1092毫米 1/8 印张：23 字数：608千字
2013年6月第一版 2013年6月第一次印刷
定价：298.00元
ISBN 978-7-112-15370-1
（23468）